W0246405

ASHGATE HANDBOOK OF PESTICIDES
AND AGRICULTURAL CHEMICALS

Ashgate Handbook of Pesticides and Agricultural Chemicals

Edited by

G W A Milne

Routledge
Taylor & Francis Group

LONDON AND NEW YORK

First published 2000 by Ashgate Publishing

Reissued 2018 by Routledge
2 Park Square, Milton Park, Abingdon, Oxon, OX14 4RN
711 Third Avenue, New York, NY 10017, USA

Routledge is an imprint of the Taylor & Francis Group, an informa business

Copyright © Taylor & Francis 2000

All rights reserved. No part of this book may be reprinted or reproduced or utilised in any form or by any electronic, mechanical, or other means, now known or hereafter invented, including photocopying and recording, or in any information storage or retrieval system, without permission in writing from the publishers.

Notice:
Product or corporate names may be trademarks or registered trademarks, and are used only for identification and explanation without intent to infringe.

Publisher's Note
The publisher has gone to great lengths to ensure the quality of this reprint but points out that some imperfections in the original copies may be apparent.

Disclaimer
The publisher has made every effort to trace copyright holders and welcomes correspondence from those they have been unable to contact.

A Library of Congress record exists under LC control number: 00102728

ISBN 13: 978-1-138-71775-6 (hbk)
ISBN 13: 978-1-138-71774-9 (pbk)
ISBN 13: 978-1-315-19623-7 (ebk)

CONTENTS

PREFACE

The *Ashgate Handbook of Pesticides and Agricultural Chemicals* contains information on 1,813 substances, including a number of mixtures, which are used widely in the agricultural environment. This group of chemical compounds has been selected from the larger collection of industrial chemicals in *Gardner's Chemical Synonyms and Trade Names*, Eleventh Edition.

The main criterion for inclusion of a material in this book is its importance as a commercially available chemical which has agricultural uses. Almost all the records describing pure chemicals carry the appropriate Chemical Abstracts Service (CAS) Registry Number and the associated EINECS (European Inventory of Existing Commercial Chemical Substances) Number. Wherever possible, a chemical is tagged with the major American and European identification numbers. In addition, all chemicals in this edition which also appear in the Twelfth Edition of the *Merck Index* have the *Merck Index* Number provided. Details of the structure of a record are provided on pages xi and xii.

Entries, whenever possible, contain detailed information on chemical composition, functions, applications, physical properties and suppliers. The trade name entries have been obtained directly from chemical manufacturers worldwide and supplemented by a research program into other secondary sources. Verification and correction of the information is a continuous process; this book contains data available at the time of its publication.

A feature of this book is the inclusion of physical properties data for pure chemicals. Properties that have been provided as available include the melting point, boiling point, density or specific gravity, refractive index, optical rotation, ultraviolet absorption, solubility and acute toxicity.

The entries are organized into 12 categories: Acaricides, Agricultural Chemicals, Animal Feeds, Fertilizers, Fungicides, Herbicides, Insecticides, Molluscicides, Nematicides, Plant Growth Regulators, Rodenticides and Slimicides. Several chemicals may belong to more than one category, in which case multiple entries will be found. Mixtures of compounds, particularly fungicides and herbicides, are used widely and are identified by name and manufacturer. For information on the components of such mixtures, refer to the respective entries.

Proprietary Considerations

Every attempt has been made to ensure the accuracy of the information provided in the *Ashgate Handbook of Pesticides and Agricultural Chemicals*. However, the publishers cannot be held responsible for the accuracy of the information, and users are reminded that:

● The reporting of a name in this book cannot imply definitive legality in establishing proprietary usage. Questions concerning legal ownership of a particular name can be resolved by due legal process.

● A manufacturer in some countries may manufacture its product under names different from those cited here. Similarly, manufacture or marketing of a product may be licensed to a separate company in another country either under the same or a different name.

We trust that readers will find that this book contains a wealth of information which is difficult to obtain from any other source. It is the intention of the publishers to produce regularly updated editions and subsets of this compilation at suitable intervals in both printed and digital form. Companies wishing to submit new or updated material for inclusion in future editions should contact George W A Milne (address on page ix).

ACKNOWLEDGEMENTS

The Editor would like to acknowledge the research work performed by Dr Ellen Zeman, the skilled programming performed by Dr Ju-Yun Li which allowed for accurate formatting and typesetting of this book, and the production work performed by Ellen Zeman and Kay Pool.

George W A Milne
Ashgate Publishing Company
131 Main Street
Burlington
Vermont 05401
USA

Telephone: 001-802-865-7641
Fax: 001-802-865-7847
e-mail: gmilne@ashgate.com

HOW TO USE THIS BOOK

The *Ashgate Handbook of Pesticides and Agricultural Chemicals* is divided into two Parts. A brief description of each Part is given below.

Part I

This Part is divided into two sections:

1   Functional Categories
    All 12 categories, and their subdivisions, are listed alphabetically with their page number in this book.

2   Synonyms and Properties
    This section contains chemicals and synonyms presented in dictionary format. Each record is identical in structure enabling the reader to select specific information efficiently. A unique Record Number has been assigned to every record, making this book versatile and easy to use. All the Indexes in Part II list these Record Numbers enabling quick cross-referencing to Part I.

Record Structure

A typical record in this book is shown on page xii. The first line contains, in bold face, the Record Number (11) and the name of the material (Dimethoate). The second line gives, if available, the Chemical Abstracts Service (CAS) Registry Number for the compound (60-51-5), the corresponding *Merck Index* Entry Number (3269) and the European Inventory of Existing Commercial Chemical Substances (EINECS) Number (200-480-3). Such numbers always appear in the same position (left, centre or right) enabling the reader to determine which source they belong to. Whenever CAS Registry Numbers are used in the text, they are always enclosed in brackets, for example [60-51-5]. The molecular formula of the compound is provided and the next

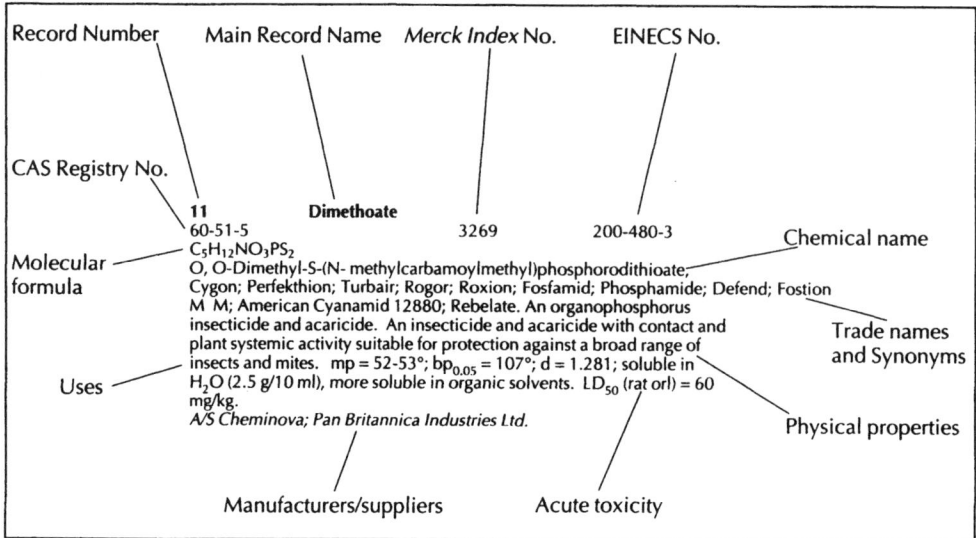

Record Number   Main Record Name   *Merck Index* No.   EINECS No.

CAS Registry No.

11   Dimethoate
60-51-5   3269   200-480-3   Chemical name
$C_5H_{12}NO_3PS_2$

Molecular formula

O, O-Dimethyl-S-(N- methylcarbamoylmethyl)phosphorodithioate;
Cygon; Perfekthion; Turbair; Rogor; Roxion; Fosfamid; Phosphamide; Defend; Fostion
M M; American Cyanamid 12880; Rebelate. An organophosphorus
insecticide and acaricide. An insecticide and acaricide with contact and
plant systemic activity suitable for protection against a broad range of
insects and mites. mp = 52-53°; $bp_{0.05}$ = 107°; d = 1.281; soluble in
H_2O (2.5 g/10 ml), more soluble in organic solvents. LD_{50} (rat orl) = 60
mg/kg.
A/S Cheminova; Pan Britannica Industries Ltd.

Uses

Trade names and Synonyms

Physical properties

Manufacturers/suppliers   Acute toxicity

line carries the chemical name of the compound. This is followed by as many as 100 synonyms, including trade names and other trivial names.

A description of the material and its known uses then follows and, when available, its physical properties are presented. These include melting point, boiling point, density or specific gravity, refractive index, optical rotation, ultraviolet absorption, solubility and acute toxicity, usually limited to oral dosage in rodents. Finally, the companies who supply, or have supplied, the product are given.

It should be noted that much of this data is available only for pure compounds. When mixtures of known composition are described, the components are given and each can be examined separately. Properties that are described, however, are those of the mixture.

Part II

This Part contains three Indexes. The purpose of each is described below:

1   CAS Registry Number Index
     This Index enables the reader to locate the Record Number for any CAS Registry Number and cross-refer to 'Synonyms and Properties' in Part I.

2   EINECS Number Index
     This Index enables the reader to locate the Record Number for any EINECS Number and cross-refer to 'Synonyms and Properties' in Part I.

3   Name and Synonym Index
     This is the master Index containing all chemical and trade names mentioned in the book, whether as a synonym or main entry. It is the most convenient place for the reader to start if only the name or synonym is known. This Index enables the reader to locate the Record Number in 'Synonyms and Properties' in Part I which relates to the main entry for that chemical.

GLOSSARY OF UNITS

Name	Description
Mass	Unless otherwise specified, mass is expressed in a multiple of grams (g), such as micrograms (μg; $= 10^{-6}$ g), milligrams (mg; $= 10^{-3}$ g), grams (g; $= 10^{0}$ g), kilograms (kg; $= 10^{+3}$ g), etc.
Volume	Volume is expressed in litres (l) or millilitres (ml) unless otherwise specified.
Temperature	When no units are cited, the temperature given is in degrees Celsius ($^{\circ}$C).
Melting point	Melting points are cited in degrees Celsius ($^{\circ}$C) unless otherwise specified.
Boiling point	When measured at atmospheric pressure, boiling points are cited with no pressure, e.g. bp $= 167^{\circ}$. At other pressures, the pressure is also cited, i.e. $bp_{0.01} = 167^{\circ}$.
Density	The measurement temperature is given as a superscript; thus a density of 1.123 measured at 25° will appear as $d^{25} = 1.123$. If the measurement was explicitly referenced to the density of water at 4°, the citation will carry both a superscript and a subscript, as in $d^{25}_{4} = 1.123$. Specific gravities are denoted by the abbreviation 'sg'.
Refractive index	Denoted by the letter n, refractive indices are usually determined at a temperature which is cited as a superscript, as in $n^{25} = 1.5432$. The wavelength of the light used in the measurement is cited as a subscript, as in $n^{25}_{546} = 1.5432$. Most commonly, the sodium D line (wavelength 549 nm) is used and in such cases, the subscript is a D, as in $n^{25}_{D} = 1.5432$.

Name	Description
Optical rotation	As with refractive indexes, optical rotations (α) are cited with the measurement temperature superscripted, and the measurement wavelength (often the sodium D line) subscripted, as in $[\alpha]_D^{25} = 105°$. When mutarotation can occur, the rotation given is an equilibrium value, measured after some time interval, which is cited, as in $[\alpha]_D^{25} = 105°(14\,hr)$.
UV absorption	The ultraviolet absorption maxima given by the material are cited in nanometers (nm = millimicrons, mμ) and the absorptivity (E, A, ε or log ε, all of which are unitless) is also given.
Acute toxicity	Wherever possible the units of toxicity are LD_{50}, i.e. the dose which is lethal to 50% of the test animals. In most cases, acute toxicity is measured with the rat, orally administered, and the result is reported as LD_{50} (rat orl) = 50 mg/kg. Other species (for example, mus = mouse; rbt = rabbit; pgn = pigeon; hmn = human; chd = child; wmn = woman; gpg = guinea pig) are occasionally cited as are other administration routes (sc = subcutaneous; ihl = inhalation; ip = intraperitoneal; iv = intravenous). Chronic toxicity data is not given.

ABBREVIATIONS

ABS	acrylonitrile-butadiene-styrene
ACE	acetylcholinesterase
ACN	acrylonitrile
alc.	alcohol
AMP	2-amino-2-methyl-1-propanol
aq.	aqueous
ASA	acrylic-styrene-acrylonitrile
BHA	butylated hydroxyanisole
BHT	butylated hydroxytoluene
BMC	bulk molding compound
bp	boiling point
BP	British Pharmacopeia
BR	butadiene rubbers, polybutadienes
B/S	butadiene/styrene
CAB	cellulose acetate butyrate
CAS	Chemical Abstracts Service
CDA	completely denatured alcohol
CI	Color Index
CMC	carboxymethylcellulose, critical micellar concentration
CNS	central nervous system
CPE	chlorinated polyethylene
CPVC	chlorinated polyvinyl chloride
CR	chloroprene rubber, polychloroprene
cs *or* cSt	centistoke(s)
CTFA	Cosmetic, Toiletries and Fragrance Association
DAP	diallyl phthalate, diammonium phosphate
DB	dichlorophenoxybutyric acid
DEA	diethanolamine, diethanolamide
DEDM	diethylol diethyl

DIBA	diisobutyl adipate
DIDA	diisodecyl adipate
DMC	4,4'-dichloro(methylbenzhydrol)
DMDM	dimethylol dimethyl
DMF	dimethlformamide
DMSO	dimethyl sulfoxide
DNPT	dinitrosopentamethylenetetramine
DOP	dioctyl phthalate
DOT	Department of Transportation
DP acid	diphenolic acid
DPG	diphenylguanidine
DTPA	diethylenetriamine pentaacetic acid
ECTFE	ethylene/chlorotrifluoroethylene copolymer
EDTA	ethylenediamine tetraacetic acid
EINECS	European Inventory of Existing Commercial Chemical Substances
EMC	electromagnetic conductive
EMI	electromagnetic interference
EO	ethylene oxide
EP	extreme pressure
EPDM	ethylene-propylene-diene rubbers
EPM	ethylene-propylene rubbers
EPR	ethylene-propylene rubber
ESCR	environmental stress crack resistance
ETFE	ethylene tetrafluoroethylene
ETU	ethylene thiourea
EVA	ethylene vinyl acetate
F	Fahrenheit
FA	fatty acid
FDA	Food & Drug Administration
FEP	fluorinated ethylene propylene
FFA	free fatty acid
FG	food grade
fp	freezing point
FRP	fiberglass-reinforced plastic(s)
GFRP	glass fiber-reinforced plastic(s)
gran.	granular, granules
GRAS	generally recognized as safe
GRP	glass-reinforced plastics, polyester
HDI	hexamethylene diisocyanate
HDL	high density lipids
HDPE	high density polyethylene
HIPS	high impact polystyrene
HLB	hydrophilic-lipophilic balance
HPLC	high performance liquid chromatography
IC	integrated circuit
ihl	inhalation
IIR	isobutylene-isoprene rubber
IPA	isopropyl alcohol
IPM	isopropyl myristate

IPP	isopropyl palmitate
IR	(synthetic) isoprene rubber, infrared
IU	international units
iv	intravenous
J	joule(s)
KTPP	potassium tripolyphosphate
LDL	low density lipids
LDPE	low density polyethylene
LED	light-emitting diode
LLDPE	linear low density polyethylene
Ltd	Limited
MA	methacrylic acid
MBCA	4,4'-methylene bis(orthochloroaniline)
MBT	mercaptobenzothiazole
MBTS	2-mercaptobenzothiazole disulfide
MCPA	(4-chloro-2-methylphenoxy) acetic acid
MDI	methylene diphenylene diisocyanate
MDM	monomethylol dimethyl
MDPE	medium density polyethylene
MEA	monoethanolamine, monoethanolamide
MEK	methyl ethyl ketone
MIBK	methyl isobutyl ketone
min.	minute(s), mineral, minimum
MIPA	monoisopropylamine, monoisopropylamide
MKP	monopotassium phosphate
MMW-HDPE	medium molecular weight high density polyethylene
MOCA	methylene bis(orthochloroaniline)
mp	melting point
MPK	methyl propyl ketone
MVTR	moisture vapor transmission rate
mw	molecular weight
N	normal
NBR	nitrile-butadiene rubber
NC	nitrocellulose
NCR	nitrile-chloroprene rubber
NEMA	National Electrical Manufacturers Association
N/F	non-flammable
NF	National Formulary
NR	natural (isoprene) rubber
NSF	National Science Foundation
NTA	nitrilotriacetic acid
OEM	original equipment manufacturer
OPP	oriented polypropylene
OTC	over the counter (i.e. non-prescription) drug
o/w	oil-in-water
Pa	Pascal
PAN	polyacrylonitrile
PBT	polybutylene terephthalate
pbw	parts by weight

PC	polycarbonate
PCA	2-pyrrolidine-5-carboxylic acid
PCP	pentachlorophenol
PCTFE	polychlorotrifluoroethylene
PE	polyethylene
PEEK	polyetheretherketone
PEG	polyethylene glycol
PEI	polyetherimide
PEK	polyetherketone
PES	polyether sulfone
PET	polyethylene terephthalate
PFA	perfluoroalkoxy
PG	polypropylene glycol
pH	hydrogen ion concentration as negative logarithm
phr	parts per hundred of rubber or resin
PIB	polyisobutylene
pK	dissociation constant as negative logarithm
PMA	phosphomolybdic acid
PMMA	polymethyl methacrylate
PO	propylene oxide
POE	polyoxyethylene, polyoxyethylated
POM	polyoxymethylene
POP	polyoxypropylene, polyoxypropylated
PP	polypropylene
PPE	polyphenylene ether
PPG	polypropylene glycol
ppm	parts per million
PPO	polyphenylene oxide
PPS	polyphenylene sulfide
PS	polystyrene
PTFE	polytetrafluoroethylene
PTMEG	polytetramethylene ether glycol
PU, PUR	polyurethane
PVA, PVAL	polyvinyl alcohol
PVAc	polyvinyl acetate
PVB	polyvinyl butyral
PVC	polyvinyl chloride
PVDC	polyvinylidene chloride
PVDF	polyvinylidene fluoride
PVE	polyvinyl ethyl ether
PVF	polyvinyl fluoride
PVM	polyvinyl methyl ether
PVP	polyvinyl pyrrolidone
RFI	radio frequency interference
RIM	reaction injection molded (molding)
RTM	resin transfer molding
RTV	room temperature vulcanizing
RV	recreational vehicle
SAN	styrene-acrylonitrile

S/B	styrene/butadiene
SBR	styrene/butadiene rubber
SBS	styrene-butadiene-styrene
SDA	specially denatured alcohol
SF	self emulsifying
SMA	styrene maleic anhydride
SMC	sheet molding compound
SPF	sun protection factor
SR	styrene rubber
SRF	semi-reinforced furnace
TBHQ	*tert*-butylhydroquinone
TDI	toluene diisocyanate
TEA	triethanolamine, triethanolamide
TFE	tetrafluoroethylene
THF	tetrahydrofuran
TIPA	triisopropanolamine
TMC	thick molding compound
TMPTA	trimethylolpropane triacrylate
TPGDA	tripropylene glycol diacrylate
TPO	thermoplastic polyolefin
UF	urea-formaldehyde
UHF	ultra-high frequency
UHMW	ultra high molecular weight
UHMWPE	ultra high molecular weight polyethylene
UL	Underwriter's Laboratory
UPVC	unplasticized polyvinyl chloride
USDA	United States Department of Agriculture
USP	United States Pharmacopeia
uv	ultra-violet
VA	vinyl acetate
VAE	vinyl acetate ethylene
VC	vinyl chloride
VdC, VDC	vinylidenechloride
VHF	very high frequency
VOC	volatile organic compounds
v/v	volume by volume
v/w	volume by weight
w/o	water in oil
w/v	weight by volume
w/w	weight by weight
XLPE	cross-linked polyethylene

PART I

MAIN ENTRIES

Functional Categories

Functional Categories

Synonyms and Properties

Acaricides

1　Abequito
50435-25-1　　　　6642
$C_9H_8ClNS_2$
4-Chloro-N-1,3-dithietan-2-ylidene-2-methylbenzeneamine.
Nimidane; Cyclic methylene (4-chloro-o-tolyl)-dithioimidocarbonate; AC 84633; ENT 29106. Acaricide. mp = 43-46°. *American Cyanamid. See* nimidane.

2　Amitraz
33089-61-1　　　　510　　　251-375-4
$C_{19}H_{23}N_3$
1,5-Di(2,4-dimethylphenyl)-3-methyl-1,3,5-triazapenta-1,4-diene.
Ovasyn; Mitac; Mitac 20; Mitaban; Baam; Triazid; Taktic; . Acaricide and insecticide for use on fruit trees and hops. *Schering Agrochemicals Ltd.; Pharmacia & Upjohn.*

3　Apollo 50C
74115-24-5　　　　2435　　　277-728-2
Clofentezine.
Suspension concentrate containing 500 g clofentezine per liter; an acaricide for use on top fruit. *Schering Agrochemicals Ltd. See* Clofentezine.

4　Azocyclotin
41083-11-8　　　　　　　255-209-1
$C_{20}H_{35}N_3Sn$
1-(Tricyclohexylstannyl)-1H-1,2,4-triazole.
BAY-BUE 1452; Peropal; tricyclotin. Acaricide used for control of mobile stages of spider mites on pome and stone fruit, grapes, citrus, vegetables, etc. mp = 218°; insoluble in H_2O (< 1 mg/l), soluble in organic solvents; LD_{50} (rat orl) = 99 mg/kg.

5　Childion
Emulsifiable concentrate of 166 g dicofol and 58.7 g tetradifon per liter; a contact acaricide. *Hortichem Ltd; ICI AgroChemicals. See* dicofol, tetradifon.

6　Cyhexatin
13121-70-5　　　　2829　　　236-049-1
$C_{18}H_{34}OSn$
Tricyclohexylhydroxystannane.
tricyclohexylstannoltricyclohexyltin hydroxide; Acarstin; Aracnol F; ENT-27395; Dowco 213; Mitacid; Plictran; TCTH; Tetran; Triran. Non-systemic acaricide with contact action. Used for control of mites on fruit crops. mp = 195-198°; insoluble in H_2O (<1 mg/l) , soluble in organic solvents; LD_{50} (rat orl) = 540 mg/kg.

7　Damfin
62610-77-9　　　　6004
Methacrifos.
Emulsifiable concentrate containing 950 g/l methacrifos; insecticide and acaricide used for pest control in stored grain. *Ciba-Geigy Agrochemicals. See* Methacrifos.

8　Demeton-S-methyl
919-86-8　　　　6129　　　213-052-6
$C_6H_{15}O_3PS_2$
S-[2-(Ethylthio)ethyl] O,O-dimethyl phosphorothioate.
methyl demeton; methylmercaptofostiol; Bay 18436; Bay 25/154; DSM; Duratox; Metasystox 55; Metasystox I; Mifatox; Persyst; Power DSM. Systemic insecticide and acaricide with contact and stomach action. Used for control of aphids and other insects in a wide variety of crops. Emulsifiable concentrate containing 580 g demeton-S-methyl per liter; a systemic organophosphorus insecticide and acaricide. bp_1 = 118°; d^{20} = 1.207; n_D^{20} = 1.5065; soluble in H_2O (3.3 g/l), more soluble in organic solvents; LD_{50} (rat orl) = 40-106 mg/kg. *Kommer-Brookwick Ltd.*

9　Dicofol
115-32-2　　　　3136　　　204-082-0
$C_{14}H_9Cl_5O$
4-Chloro-α-(4-chlorophenyl)-α-(trichloromethyl)benzenemethanol.
1,1-bis(p-chlorophenyl)-2,2,2-trichloroethanol; DTMC; ENT 23648; FW 293; Kelthane; Mitigan; Acarin. Active ingredient; dicofol; a specific acaricide and miticide. mp = 77-78°; bp = 225°; d^{10} = 1.1234; n_D = 1.1234; $[\alpha]_D$ = 100°; , insoluble in H_2O, soluble in organic solvents; λ_m = 226, 258, 266, 276 nm (logε = 4.43, 2.82, 2.85, 2.60); LD_{50} (rat, orl) = 1495 mg/kg. *Makhteshim Chemical Works Ltd.; Agan Chemical Manufacturers Ltd.*

10　Dienochlor
2227-17-0　　　　3154　　　218-763-5
$C_{10}Cl_{10}$
1,1',2,2',3,3',4,4',5,5'-Decachlorobi-2,4-cyclopentadien-1-yl.
HRS-16; Pentac. Acaricide used to control mites on ornamental plants. mp = 121-122°; λ_m = 330 nm (ε 2950).

11　Dimethoate
60-51-5　　　　3269　　　200-480-3
$C_5H_{12}NO_3PS_2$
O,O-Dimethyl-S-(N-methylcarbamoylmethyl) phosphorodithioate.
Cygon; Perfekthion; Turbair; Rogor; Roxion; Fosfamid; Phosphamide; Defend; Fostion M M; American Cyanamid 12880; Rebelate. An organophosphorus insecticide and acaricide. An

insecticide and acaricide with contact and plant systemic activity suitable for protection against a broad range of insects and mites. mp = 52-53°; $bp_{0.05}$ = 107°; d = 1.281; soluble in H_2O (2.5 g/10 ml), more soluble in organic solvents; LD_{50} (rat orl) = 60 mg/kg. *A/S Cheminova; Pan Britannica Industries Ltd.*

12   Divipan
62-73-7   3129   200-547-7
Dichlorvos.
Active ingredient: dichlorvos; one of the most useful fast-acting agricultural insecticides-acaricides. *Makhteshim Chemical Works Ltd.*

13   Fenpropathrin
39515-41-8   4033   254-485-0
$C_{22}H_{23}NO_3$
Cyano(3-phenoxyphenyl)methyl 2,2,3,3-tetramethylcyclopropanecarboxylate.
fenpropathrine; Danitol; Herald; Kilumal; Meothrin; Ortho Danitol; Rody; S-3206; Tame. A pyrethroid-based acaricide and insecticide with contact, stomach and repellent action. Used for control of mites and insects in fruit and vegetable crops. mp = 45-50°; d^{25} = 1.15; almost insoluble in H_2O (0.33 mg/l), more soluble in organic solvents; LD_{50} (rat orl) = 71 mg/kg. *Shell UK.*

14   Harle's solution
7784-46-5   8721   232-070-5
Sodium arsenite.
A solution of sodium arsenite. Used as a topical acaricide. *See* sodium arsenite.

15   Hortichem Spraying Oil
Emulsifiable concentrate containing 710 g/l petroleum oil; an insecticide and acaricide. *Hortichem Ltd.*

16   Malathion
121-75-5   5740   204-497-7
$C_{10}H_{19}O_6PS_2$
Diethyl [(dimethoxyphosphinothioyl)thio]butanedioate.
Maldison; Insecticide No. 4049; Carbofos; Mercaptothion; Phosphothion; Cythion; Chemathion; Carbophos; Emmatos; Fosfothion; Fyfanon; Karbofos; Kop-Thion; Malacide; Malagran; Malamar; MLT; Sadofos; Calmathion; Carbetox; Carbethoxy Malathion; Carbetovur; Celthion; Cimexan; Compound 4049; Detmol MA; Malaspray; Ethiolacar; Etiol; Cleensheen; Lice Rid. Cholinesterase inhibitor, behaves as a non-systemic acaricide and insecticide with contact, stomach and respiratory action. Used for broad spectrum control of sucking and chewing insects and mites in a wide variety of crops. Particularly useful on Mediterranean fruit fly. mp = 2°; $bp_{0.7}$ =

156-157°; sg^{25} = 1.23; n_D^{25} = 1.4985; soluble in H_2O (145 mg/l), more soluble in organic solvents; LD_{50} (rat orl) = 1375-2800 mg/kg. *Allchem Industries; Am. Cyanamid; Sariaf SpA.*

17   Mesurol®
2032-65-7   6050   217-991-2
Methiocarb.
Versatile product formulated for different uses; especially as a molluscicide against slugs and snails as well as a seed dressing for repelling depredating birds; also as insecticide/acaricide against foliar-feeding caterpillars and sucking pests on various crops. *Bayer AG. See* methiocarb.

18   Methamidophos
10265-92-6   6014   233-606-0
$C_2H_8NO_2PS$
O,S-Dimethyl phosphoramidothioate.
acephate-met; ENT-27396; Bay 71628; Filitox; Monitor; Patrole; Pillaron; SRA 5172; Tam; Tamanox; Tamaron. Systemic insecticide and acaricide with contact and stomach action. Absorbed by roots and leaves. Cholinesterase inhibitor. For control of chewing and sucking insects and spider mites in vegetable and fruit crops. mp = 46°; sg_{20} = 1.31; n_D^{40} = 1.5092; soluble in H_2O (>2 kg/l), less soluble in organic solvents; LD_{50} (rat orl) = 20 mg/kg.

19   Methidathion
950-37-8   6048   213-449-4
$C_6H_{11}N_2O_4PS_3$
Phosphorodithioic acidS-[(5-methoxy-2-oxo-1,3,4-thiadiazol-3(2H)-yl)methyl O,O-dimethyl ester. GS-13005; Supracide; Ultracid. Insecticide, acaricide. mp = 39-40°; soluble in H_2O (<1%), soluble in organic solvents; LD_{50} (rat orl) = 31 mg/kg.

20   Methomyl
16752-77-5   6062   240-815-0
$C_6H_{12}N_2O_3S$
S-Methyl-N[(methylcarbamoyl)oxy]-thioacetimidate.
Nudrin; Lannate; Lannate(R); Lanox; Methomyl 5G; Lannabait; Lannate LB; Insecticide 1179; Lanox 216; LANOX 90; Mesomile; Nu-bait II; Thiobutan-2-one, O-(methylcarbamoyl)oxime; Flytek; Kipsin; Dupont 1179; Memilene; Methavin; Methomex; Nudrin. A cholinesterase inhibitor used as a systemic insecticide and acaricide. Used for control of a wide range of insects and spider mites in many fruits and vegetables. Also for control of flies in animal and poultry houses and in dairies. mp = 78-79°; d_4^{24} = 1.2946; soluble in H_2O (5.8 g/ml), more soluble in organic solvents; LD_{50} (rat orl) = 17 mg/kg. *Makhteshim Chemical Works Ltd.*

21 Murfite
Acaricide. Murphy Chemical Co Ltd.

22 Omethoate
1113-02-6   214-197-8
$C_5H_{12}NO_4PS$
O,O-Dimethyl S-[2-(methylamino)-2-oxoethyl] phosphorothioate.
dimethoate-met; Bay 45432; Folimat®; S-6876. Systemic insecticide and acaricide with contact and stomach action. Used for control of spider mites, aphids, beetles, caterpillars, scale insects, thrips, suckers, fruit flies etc. in fruit and vegetable crops and in forestry. dec 135°; $d^{20} = 1.32$; $n_D^{20} = 1.4987$; soluble in H_2O, organic solvents; LD_{50} (rat orl) = 50 mg/kg. *Bayer AG; Bayer plc.*

23 Parathion
56-38-2   7167   200-271-7
$C_{10}H_{14}NO_5PS$
O,O-Diethyl O-*p*-nitrophenyl phosphorothioate.
diethoxy, nitro-phenoxy phosphorothioate; DNTP; S.N.P.; E-605; AC-3422; ENT-15108; Alkron; Alleron; Aphamite; Etilon; Folidol®; Fosferno; Niran; Paraphos; Rodiatox; Thiophos. Powerful insecticide, acaricide. mp = 6°; bp = 375°; $d_4^{25} = 1.26$; $n_D^{25} = 1.5370$; insoluble in H_2O, soluble in organic solvents; LD_{50} (rat orl) = 13 mg/kg. *Bayer AG.*

24 Parathion-methyl
298-00-0   6183   206-050-1
$C_8H_{10}NO_5PS$
O,O-Dimethyl O-(4-nitrophenyl) phosphorothioate.
methyl parathion; metaphos; OMS 213; ENT 17292; Bladan M; Cekumethion; Devithion; Folidol-M; Fulkil; Metacide; Methyl-bladan; Nitrox; Parataf; Paratox; Partron M; Penncap-M; Tekwaisa; Wofatox. Cholinesterase inhibitor which serves as a non-systemic insecticide and acaricide with contact and stomach action. Used for control of sucking and chewing insects and mites in a wide range of crops. Insecticide and acaricide used extensively in cotton producing areas. mp = 35-36°; $bp_1 = 154°$; soluble in H_2O (55 mg/l), more soluble in organic solvents; LD_{50} (rat orl)= 14-24 mg/kg. *Bayer AG; A/S Cheminova.*

25 Phosalone
2310-17-0   7489   218-996-2
$C_{12}H_{15}ClNO_4PS_2$
S-[(6-Chloro-2-oxo-3(2H)-benzoxazolyl)methyl] O,O-diethylphosphorodithioate.
ENT 27163; 11974 RP; Azofene; Rubitox; Zolone. Non-systemic insecticide and acaricide with contact and stomach action. Used for control of sucking and chewing insects, spider mites, Colorado beetles, aphids, bollwormsand stem borers. mp = 45-48°; soluble in H_2O (10 mg/l), more soluble in organic solvents; LD_{50} (rat orl) = 120-175 mg/kg.

26 Pirimicarb
23103-98-2   7651   245-430-1
$C_{11}H_{18}N_4O_2$
2-(Dimethylamino)-5,6-dimethyl-4-pyrimidinyl dimethylcarbamate.
Power Demo; Pirimor; Abol; Aficida; Aphox; Fernos; Rapid. Granules containing 50% w/w pirimicarb; for control of aphids. mp = 90.5°; soluble in H_2O (0.27 g/100 ml), soluble in most organic solvents; LD_{50} (frat orl) = 147 mg/kg. *Kommer-Brookwick Ltd..*

27 Pyrimithate
5221-49-8   8172   226-020-1
$C_{11}H_{20}N_3O_3PS$
Phosphorothioic acid O-[2-(dimethylamino)-6-methyl-4-pyrimidinyl] O,O-diethyl ester.
ICI-29661; Diothyl; pyrimitate. Used as an acaricide and an insecticide. $bp_{0.04} = 128-132°$; d= 1.165; insoluble in H_2O, soluble in organic solvents. *ICI Agrochemicals.*

28 Quinomethionate
2439-01-2   7113   219-455-3
$C_{10}H_6N_2OS_2$
6-Methyl-1,3-dithiolo[4,5-b]quinoxalin-2-one.
chinomethionat; oxythioquinox; quinoxalines; ENT 25606; Morestan®; Bay 36205; Chinomethionate; Forstan; SS 2074. A wettable powder containing 25% w/w quinomethionate; fungicide with protective and eradicative action against powdery mildews on pome, stone and small fruits, cucurbits and ornamentals; as acaricide effective against eggs and mobile stages of mites and mp = 169-170°; soluble in H_2O (1 mg/l), more soluble in organic solvents; LD_{50} (rat orl) = 2500-3000 mg/kg. *Bayer AG.*

29 Strobane
Toxaphene.
terpene polychlorinates; Dichloricide Aerosol; Dichloricide Mothproofer; Insecticide 3960-X14. A trademark for an insecticide and acaricide; it is based on mixed polychlorinated terpene isomers and contains 66% chlorine. *See* toxaphene.

30 SuXon
Insecticide, acaricide. Incitec Ltd. *See* chlorpyrifos.

31 Tedion V-18
116-29-0   9339   204-134-2
$C_{12}H_6Cl_4O_2S$
1,2,4-Trichloro-5-[(4-chlorophenyl)sulfonyl]-benzene.
4-chlorophenyl-2,4,5-trichlorophenyl sulfone;

2,4,4',5-tetrachlorodiphenyl sulfone; p-chlorophenyl 2,4,5-trichlorophenyl sulfone; Acaroil TD; Acarvin; Agrex T-7.5; Aracnol K; Mitifon; Tetranol; V-18. Tetradifon; emulsifiable concentrate containing 125 g propiconazole, 350 g tridemorph per liter; non-systemic selective acaricide for use against mite infestation in orchards, citrus fruit plantations, hop fields, groundnut plantations, vegetable plots, cotton fields and on ornamental plants; red spider mite control in horticultural crops. mp = 148-149°; d_{20} = 1.515; soluble in H_2O 0.08 mg/l), more soluble in organic solvents. *Duphar BV; Hortichem Ltd.* See propiconazole.

32   Thiodan
115-29-7   3614   204-079-4
$C_9H_6Cl_6O_3S$
6,7,8,9,10,10-Hexachloro-1,5,5a,6,9,9a-hexahydro-6,9-methano-2,4,3-benzodioxathiepin 3-oxide.
Endosulfan; benzoepin; OMS 750; ENT 23979; Beosit; Cyclodan; Chlortiepin; Devisulphan; Endocel; Endosol; FMC 5462; Hilda; Hoe 2671; Insectophene; Malix; Rasayansulfan; Thifor; Thimul; Thionex; Thiosulfan. Non-systemic in-secticide and acaricide with contact and stomach action. Controls sucking, chewing and boring insects and mites on a wide variety of crops. The commercial product is a mixture of an α-isomer (mp = 108-110°) and a β-isomer (mp = 208-210°). mp = 109°; $bp_{0.7}$ = 106°; d_{20} = 1.745; soluble in H_2O (0.32 mg/l), more soluble in organic solvents; LD_{50} (rat orl)= 70 mg/kg. *Hoechst UK.*

33   Thiometon
640-15-3   211-362-6
$C_6H_{15}O_2PS_3$
S-[2-(Ethylthio)ethyl] O,O-dimethyl phosphorodithioate.
S-2-ethylthioethyl O,O-dimethyl phosphorodi-thioate; dithiometon; M-81; Bay 23129; Ekatin; Medrin; nimeton. Systemic insecticide and acaricide with contact and stomach action. Cholinesterase inhibitor, used for control of sucking insects in fruit and vegetable crops. $bp_{0.1}$ = 110°; d_{20} = 1.209; n_D^{20} = 1.5515; soluble in H_2O (200 mg/l), more soluble in organic solvents; LD_{50} (rat orl) = 125 mg/kg. *Sandoz.*

34   Sulfur
7704-34-9   9142   231-722-6
S
Sulfur.
soufre; Alfa, Aquilite; Cosan; Elosal; Golden Dew; Imber, Kolodust; Kolofog; Kolospray; Kumulus; Magnetic 6; Solfa; Suffa; Sulfex; Sulflox; Sulfospor; Sulphotox; Super Six; That; Thiolux; Thion; Thiovit; This; Tiolene; Uniflow; Zolvis. Protectant

fungicide for control of scab in fruits and mildews on a variety of crops. Acaricide. mp = 113°, 114° or 119°; bp = 444°; d = 2.07; insoluble in H_2O, slightly soluble in organic solvents; non-toxic to humans, animals. *Pan Britannica Industries Ltd.*

35   Torque
13356-08-6   4004   236-407-7
$C_{60}H_{78}OSn_2$
Hexakis(2-methyl-2-phenylpropyl)distannoxane. bis[tris(2-methyl-2-phenylpropyl)tin]   oxide; Fenbutatin oxide; fenbutatin oxyde; hexakis; ENT 27738; Osadan; SD 14114; Vendex. A non-systemic acaricide with contct and stomach action. Used for control of phytophagous mites in fruit crops. mp = 138-139°; poorly soluble in H_2O (0.005 mg/l), more soluble in organic solvents; LD_{50} (rat orl) = 2631 mg/kg. *ICI Agrochemicals.*

36   Triazophos
24017-47-8   9736   245-986-5
$C_{12}H_{16}N_3O_3PS$
O,O-Diethyl   O-(1-phenyl-1H-1,2,4-triazol-3-yl) phosphorothioate.
HOE   2960;   Hostathion.   Broad   spectrum insecticide and acaricide. Used for control of aphids and insects in a wide variety of crops. mp = 2-5°; bp dec; d^{20} = 1.247; n_D^{20} = 1.5501; soluble in H_2O (30 mg/l), more soluble in organic solvents; LD_{50} (rat orl) = 57 mg/kg.

37   Vydate®
23135-22-0   245-445-3
$C_7H_{13}N_3O_3S$
2-(Dimethylamino)-N-[[(methylamino)carbonyl]-oxy]-2-oxoethanimidothioic acid methyl ester.
oxamyl;   N',N'-dimethyl-N-[(methylcarbamoyl)-oxy]-1-thiooxamimidic acid methyl ester; N,N-dimethyl-α-methylcarbamoyloxyimino-α-(methyl-thio)acetamide; methyl 1-(dimethylcarbamoyl)-N-(methylcarbamoyloxy)thioformimidate; thioxamyl; DPX-1410; Blade; DPX 1410. Granules containing 10% w/w oxamyl; contact and systemic insecticide,   acaricide   and   nematicide. Cholinesterase inhibitor. mp = 108-10°; d = 0.97; soluble in H_2O (280 g/l), more soluble in organic solvents; LD_{50} (rat orl) = 5.4 mg/kg. *DuPont UK.*

Agricultural Chemicals

38   Betrox
7647-14-5   8742   231-598-3
ClNa
Sodium chloride.
Sodium chloride fertilizer. *ICI Chem & Polymers Ltd.* See sodium chloride.

39   Bilt-Cote®

1332-58-7   5294   296-473-8
Kaolin.
Kaolinite; China clay; Bolus alba; Porcelain clay; Aluminum silicate hydroxide; Kaopectate; Aluminum silicate (hydrated); Aluminum silicate dihydrate. Kaolin clay; used for agricultural prill conditioning. *R. T. Vanderbilt Co Inc. See* Kaolin.

40   Birlane

470-90-6   2137   207-432-0
$C_{12}H_{14}Cl_3O_4P$
Chlorfenvinphos.
2-Chloro-1-(2,4-Dichlorophenyl)ethenyl phosphoric acid, diethyl ester; Dermaton; CFVP; CVP; Sapecron; Steladone; Supona; O,O-diethyl-O-1-(2',4'-dichloro-phenyl)-2-chlorovinyl phosphate; chlorofenvinphos; phosphoric acid 2-chloro-1-(2,4-dichlorophenyl)ethenyl diethyl ester; Apachlor; 2-chloro-1-(2,4-dichlorophenyl)ethenyl diethyl phosphate; Compound 4072; SD 7859; Phosphoric acid, 2-chloro-1-(2,4-dichlorophenyl) vinyl diethyl ester. Liquid seed dressing containing chlorfenvinphos for winter wheat. *ICI Chem & Polymers Ltd. See* chlorfenvinphos.

41   Bombardier

1897-45-6   2219   217-588-1
$C_8Cl_4N_2$
Chlorothalonil.
Forturf; Bravo; Exotherm; m-TCPN; Sweep; TCIN; Termil; TPN; Daconil; 2,4,5,6-tetrachloro-1,3-benzenedicarbonitrile; tetrachloroisophthalo-nitrile; 2,4,5,6-tetrachloro-1,3-dicyanobenzene; Evade; 1,3-dicyano-2,4,5,6-tetrachlorobenzene; DAC-2787; Daconil 2787; Exotherm Termil; Chlorothanonil; Farber; Jupital; Ole; Pillarich; Repulse; Taloberg; Tuffcide; Black Leaf Lawn & Garden Fungicide; Bonide; ClortoCaffaro; Clortosip; Dexol Fungicide Containing Daconil; Dragon Daconil 2787; Ferti-lome; Green Charm Multi-Purpose Fungicide; Green Thumb Lawn & Garden Fungicide; Ortho Multi-Purpose Fungicide Daconil 2787; Pennington's Pride Multi-Purpose Fungicide; Pro-Care Multi-Purpose Fungicide; Rigo's Best Lawn & Garden Fungicide; SA Lawn Ornamental & Vegetable Fungicide; Security Fungi-Gard; 2,4,5,6-tetrachloroisophthalonitrile; bravo-w-75; chloro-alonil; Dacobre; Echo 75; Vanox. Fungicide for a wide range of agricultural crops. *Farm Protection Ltd. See* Chlorothalonil.

42   Bromodan

1715-40-8   216-996-7
$C_8H_5BrCl_6$
Bromomethyl-1,2,3,4,7,7-hexachloro-2-norbornene.
Alugan. A brominating agent. *Fisons plc, Horticultural Div. See* bromocyclene.

43   Bubber Shet

57-13-6   10005   200-315-5
CH_4N_2O
Carbonyl diamine.
Superprill; Bubber Shet; Carbamide; Carbonyl diamine; Carbamimidic acid; Isourea; aquadrate; ureaphil; ureophil; Aquacare/HP; Nutriplus; Urecare; Urederm. Urea (prilled); fertilizer. Prilled urea containing minimum of 46% available nitrogen; used as a fertilizer to supply nitrogen to the crops for better yield per acre. *Columbia Nitrogen Corporation; Dawood Hercules Chemicals Ltd. See* Urea.

44   Clopyralid

1702-17-6   2462   216-935-4
$C_6H_3Cl_2NO_2$
3,6-Dichloro-2-pyridinecarboxylic acid.
Lontrel; Stinger; Dowco 290; 3,6-dichloro-2-picolinic acid; 3,6-dichloropicolinic acid; Lontril F; Lontril T; 3,6-dichloro-pyridinecarboxylic acid; Hadranol. Plant growth regulator. *Makhteshim Chemical Works Ltd. See* Clopyralid.

45   Elliott's Lawn Sand

10028-22-5   4079   233-072-9
$Fe_2O_{12}S_3$
Ferric sulfate.
Iron (III) sulfate; ferric persulfate; ferric sesquisulfate; ferric tersulfate; Sulfuric Acid, Iron(3+) Salt (3:2); Iron Tersulfate; Ferric sulfate monohydrate; Elliott's Moss Killer. Used for moss control in turf. *Thomas Elliott Ltd. See* Ferric Sulfate.

46   Elocril

18181-70-9   5053   242-069-1
$C_8H_8Cl_2IO_3PS$
O-(2,5-Dichloro-4-iodophenyl) O,O-dimethyl phosphorothioate.
Iodofenphos; Nuvanol-N; jodfenphos; Alfacron; Iodofenfos. An organophosphorus insecticide. *Ciba-Geigy Agrochemicals. See* Iodofenphos.

47   Ethyl Hexanediol

94-96-2   3790   202-377-9
$C_8H_{18}O_2$
2-Ethyl-1,3-hexanediol.
ethyl hexylene glycol; ethyl hexanediol; Rutgers 6-12; Carbide 6-12; 3-hydroxymethyl-n-heptan-4-ol; EH diol; 2-ethylhexane-diol-1,3; 2-Ethyl-1,2-Hexanediol; Ethyl-1,3-Hexane Diol-2; 2-ethyl hexanediol; Ethyl hexylene glycol; ethohexadiol; 2-ethyl-1,3-hexylene glycol; 2-ethyl-2-propyl-1,3-propanediol; octylene glycol; 6-12 insect repellent; EHD; Ethyl-1,3-hexanediol; Ethyl-3-propyl-1,3-propanediol; Hydroxymethyl-n-heptan-4-ol; Insect repellant; Repellent 6-12; 2-Ethylhexane-1,3-diol. Insect repellent, cosmetics, vehicle and solvent in printing inks, medicine,

chelating agent for boric acid. mp = -40; bp = 241-249°; d = 0.9330; n_D^{20} = 1.4497; soluble in H_2O, ethanol, isopropanol, propylene glycol, castor oil; LD_{50} (rat orl) = 1400 mg/kg. Hûls Am.; Union Carbide.

48   Folpet
133-07-3   4255   205-088-6

$C_9H_4Cl_3NO_2S$
2-[(Trichloromethyl)thio]-1H-isoindole-1,3(2H)-dione.
Folpan; N-(trichloromethylthio)phthalimide; N-(trichloromethylmercapto)phthalimide; Phaltan; Phalton; Folpel; Folpex; Trichloromethyl(thio)-phthalimide; Folpet; 2-[(trichloromethyl)thio]-1H-Isoindole-1,3(2H)-dione; Folpan; Phalton; Phaltan; Folpel; Folpex. Agricultural fungicide. Used for control of downy mildews, powdery mildews, leaf spot diseases, scab, etc. in fruit and vegetable crops. mp = 177°; insoluble in H_2O, slightly soluble in organic solvents; LD_{50} (rat orl) > 10000 mg/kg. Makhteshim Chemical Works Ltd.

49   Garden Lime
1317-65-3   5515   215-279-6

$CCaO_3$
Calcium carbonate (chalk).
Limestone; Marble Chips. Ground limestone; used to increase pH of acid soils. Vitax Ltd. See Limestone.

50   Halobrom
126-06-7   204-766-9

$C_5H_6BrClN_2O_2$
1-Bromo-3-chloro-5,5-dimethylhydantoin.
BCDMH; Bromo-1-chloro-5,5-dimethyl-2,4-imid-azolinedione; Bromo-1-chloro-5,5-dimethyl-hy-dantoin; Imidazolidinedione, 3-bromo-1-chloro-5,5-dimethyl-. Broad spectrum biocide for control of algae, bacterial and fungal slimes in swimming pools and industrial water systems; nonflammable. Dead Sea Bromine.

51   Kankerex
21908-53-2   5936   244-654-7

HgO
Mercuric oxide.
Mercury(II)oxide, red; Mercury oxide; Santar; Yellow oxide of mercury; Mercury(II)oxide, yellow. Control of canker in apples and pears. dec 500°; d = 11.14. Universal Crop Protection Ltd.

52   Kloben®
555-37-3   6523   209-096-0

$C_{12}H_{16}Cl_2N_2O$
N-Butyl-N'-(3,4-dichlorophenyl)-N-methylurea.
3-(3,4-dichlorophenyl)-1-methyl-1-n-butylurea; Kloben Neburon; Granurex; Neburex; Neburon; Herbalt; Kloben; N-butyl-N'-(3,4-dichlorophenyl)-N-methylurea; 1-Butyl-3-(3,4-dichlorophenyl)-1-methylurea; 1-Butyl-N'-(3,4-dichlorophenyl)-N-methylurea. For the agriculture industry. mp = 101.5-103°. DuPont UK.

53   Kumulus® DF, FL
7704-34-9   9142   231-722-6

S
Sulfur.
Sulphur; brimstone. For control of diseases and spider mites in fruit, vines, vegetable, ornamentals, and agricultural crops. mp = 117-120°; bp = 444°; d = 2.060. BASF AG.

54   Lime Nitrate
10124-37-5   1729   233-332-1

$CaN_2O_6.4H_2O$
Calcium nitrate.
Lime nitrate; Nitric Acid, Calcium Salt; Lime Saltpeter; Norwegian Saltpeter; Nitrocalcite; Calcium Nitrate-15N2. (Lime saltpeter). Fertilizer.

55   Lime Nitrogen
156-62-7   1702   205-861-8

$CaCN_2$
Calcium cyanamide.
Calcium Carbamide; Lime Nitrogen; Alzodex; Cyanamide, Calcium Salt (1:1); Calcium Carbimide; Alzodef; Cyanamide, calcium; nitro-lime; Cyanamid pam-amd; aero-cyanamid; nitrogen lime; cy-l 500; Temposil. Fertilizer. Used in nitrogen products, pesticide, hardening iron, steel.

56   Maneb
12427-38-2   5761   235-654-8

$C_4H_6MnN_2S_4$
[[1,2-Ethanediylbis[carbamodithioato]](2-)] manganese.
Manganese, [ethylenebis[dithiocarbamato]]-; m-Diphar; Amangan; Carbamic acid, ethylene-bis[dithio-, manganese salt; Chem neb; Chloroble B; CR 3029; Dithane M 22; Dithane M 22 special; Dithane M-45; Dithane S-31; Ethylene bis[dithiocarbamic acid], manganese salt; EBDC, manganese salt; F 10; Griffin Manex; Kypman 80; Labilite; Lonocol M; M-Diphar; Manam; 249; Maneb 80; Maneb-R; Maneba; manebe; MEB; ENT 14875;Kypman; Man-Zox; Manex; Manox; Manzi; Polyram M; Trimangol. Fungicide with protective action. Used in control of many fungal diseases, e.g. blight, leaf spot, rust, downy mildew, scab etc. mp = 192-204°; insoluble in H_2O and common solvents; LD_{50} (rat orl) = 3000 mg/kg. Rohm & Haas; Cumberland; Crystal; Drexel; BASF; Chiltern Farm Chemicals Ltd; Dupont; Pennwalt Holland.

57   Mild Lime
471-34-1   1697   207-439-9
CCaO₃

Calcium carbonate (chalk).
Limestone; Marble Chips. Is known in agriculture as mild lime. *See* calcium carbonate.

58   Sterilite Hop Defoliant
120-12-7   721   204-371-1
C₁₄H₁₀

Anthracene.
Paranaphthalene; anthracin; green oil; tetra olive n2g. Anthracene oil; used for chemical stripping in hop vines. *Coventry Chemicals Ltd.*

Animal Repellents

59   Hoppit
8208

Bitter wood; Bitter ash. Quassia; animal repellent for outdoor crops. Has been used as an anthelmintic. *Fieldspray.*

Animal Dips

60   Coppertox
Livestock dips and sprays. *The Wellcome Foundation Ltd.*

Fertilizers

61   Blood Meal
A nitrogenous fertilizer prepared by coagulating blood, drying and grinding the product; contains on average, from 11-14% nitrogen, and 0.75% phosphorus.

62   Borrechel
Lignosulfonate containing chelated trace elements; used as micronutrients. Borregaard Ligno Tech.

63   Bygran F
Tecnazene and iodophor as dry granules; for controlling sprouting, dry rot and other storage diseases in potatoes. *Wheatley Chemical Co Ltd.* *See* tecnazene.

64   Floranid® N 32
Special slow-release nitrogenous fertilizer with 32% nitrogen for intensive horticultural crops, ornamentals, and lawns. *BASF AG.*

65   Hydro
Fertilizers, chemicals, gases and plastics. Norsk Hydro AS.

66   Leuna Saltpeter
A double salt of ammonium sulfate and nitrate; a fertilizer similar to Chilean nitrate in its action.

67   Leunaphos
A mixture of phosphate, nitrate, and sulfate of ammonia. fertilizer.

68   Librebor®
An organo-boron complex for correcting boron deficiency in most crops. *Allied Colloids Ltd.*

69   Limbux®
A form of mechanically slaked lime; used in agriculture, building and construction, metallurgical and chemical industries. *ICI Chem & Polymers Ltd.*

70   Liquid Feed for Hanging Baskets
Liquid concentrate containing NPK 4:2:6 plus trace elements; liquid fertilizer for hanging baskets, tubs, planters and window boxes. *Vitax Ltd.*

71   Liquid Growmore
Liquid concentrate containing NPK 7:7:7: general purpose liquid fertilizer. *Vitax Ltd.*

72   Liquid Q4 Borders and Beds Fertiliser
Liquid concentrate containing NPK 5.3:5.3:10 plus trace elements; border and bedding plant fertilizer. *Vitax Ltd.*

73   Liquid Tomato Feed
Liquid concentrate containing NPK 4.5:4.5:9 plus Mg; fertilizer. *Vitax Ltd.*

74   Liquinure
Liquid fertilizer. Fisons plc, Horticultural Div.

75   Ortho-Gro
Liquid plant food. *Monsanto (Solaris).*

76   Pan Scale
The calcium sulfate, containing some sodium chloride, which settles out during the crystallization of salt from brine. It is sold as salt lick for cattle, also for manuring purposes.

Feed Additives

77   Forociben Premix
Sulfonamide animal feed additive. *Ciba plc.*

Fungicides

78   Lesan
Fungicide; used for control of soil some diseases on ornamentals. *Bayer AG.*

Plant Growth Regulators

79   Arotex
Growth regulator containing 644 g chlormequat and 32.2 g choline chloride per liter; for use on wheat, oats or rye. *ICI Agrochemicals.* See chlormequat, choline chloride.

80   Ashlade 5C
Mixture of chlormequat and choline chloride; plant growth regulator. *Ashlade Formulations Ltd.* See chlormequat and choline chloride.

81   Harmony®
Metsulfuron + methyl + thifensulfuron + methyl. Mixture of metsulfuron-methyl and thifensulfuron-methyl; for control of annual dicotyledons in cereals. *Du Pont UK.* See metsulfuron-methyl, thifensulfuron-methyl.

82   Keriroot
Hormone rooting powder. *ICI Chem & Polymers Ltd.*

83   Limit
N-(Acetylamino)methyl-2-chlor-N-2,6-diethyl-phenylacetamide.
Turf grass regulator. *Monsanto Co.*

84   Turbo-Grass
Mineral and plant extracts in a water base containing cytokinin, B-vitamin, morphogenic and porphyrin activity to aid in increased plant metabolism and yield; for all agricultural, horticultural, and forestry products. *SN Corp/Appropriate Technology Ltd.*

Herbicides

85   Hebron Pabracr
Cetrimide + chlorpropham.
A suspension concentrate containing 80 g cetrimide and 80 g chlorpropham per liter. Soil-acting herbicide for lettuce. *Atlas Interlates Ltd.* See cetrimide, chlorpropham.

86   Hedonal®
2,4-D MCPA + dichlorprop + MCPP.
Range of herbicides containing 2,4-D MCPA, dichlorprop, MCPP either alone or in combinations; growth regulator herbicide used for control of weeds in cereals. Bayer AG. See 2,4-D MCPA, dichlorprop, MCPP.

Metal-Containing Fertilizers

87   Librel®
Chelated metallic micronutrients designed for soil or foliar application to correct specific trace element deficiencies in crops. *Allied Colloids Ltd.*

88   Libreleaf®
Copper, iron and manganese ligno-sulfonate chelates for correcting specific trace element deficiencies; for foliar application only. *Allied Colloids Ltd.*

89   Libspray®
A range of complete foliar fertilizers containing chelated trace elements in powder and liquid form to supplement soil applied fertilizers. *Allied Colloids Ltd.*

Seed Preservatives

90   Aspulum
Mercury derivative of chlorophenol; seed preservative.

91   Fusariol
Mercury-formaldehyde preparation; seed preservative.

92   Harvesan
Mercurial seed dressing. *The Boots Co plc.*

93   Lindex
Insecticide and fungicide seed dressing. *DowElanco Ltd.*

Miscellaneous Agricultural Chemicals

94   Acquit®
For agriculture industry. *DuPont UK.*

95   Alisol®
For agriculture industry. *DuPont UK.*

96   Avaunt®
For agriculture industry. *DuPont UK.*

97   Balance®
For agriculture industry. *DuPont UK.*

98   Besiege®
For agriculture industry. *DuPont UK.*

99   Bi-play®
For agriculture industry. DuPont UK.

100   Bozzle
Container for agrochemicals; used with the Electrodyn sprayer. *ICI Chem & Polymers Ltd.*

101 Bullion®
For agriculture industry. *DuPont UK.*

102 Calibre®
For agriculture industry. *DuPont UK.*

103 Cameo®
For agricultural industry. *DuPont UK.*

104 Charisma®
For the agriculture industry. *DuPont UK.*

105 Clarion®
For the agriculture industry. *DuPont UK.*

106 Colstar®
For the agriculture industry. *DuPont UK.*

107 Curasol
Soil erosion inhibitor. *Hoechst UK.*

108 Curbetan®
For the agriculture industry. *DuPont UK.*

109 Debut®
For the agriculture industry. *DuPont UK.*

110 Drift Proof
Non-phytotoxic drift control agent, spreadersticker and pesticide deposit builder. *W A Cleary.*

111 Du Pont Enrich®
For the agriculture industry. *DuPont UK.*

112 Duet®
For the agriculture industry. *DuPont UK.*

113 Edit®
For the agriculture industry. *DuPont UK.*

114 Enhance®
For the agriculture industry. *DuPont UK.*

115 Fielder®
For the agriculture industry. *DuPont UK.*

116 Finish®
For the agriculture industry. *DuPont UK.*

117 Fl-Mo 80/20
Modified alcohol ethoxylate.
Agricultural adjuvant. *Witco/Organics.*

118 Folicote Transpiration Minimizer
Refined wax, emulsifiers, preservatives, minimum 50% solids; FDA approved for use on edible crops; reduces water loss from plant foliage, winter protection, transplanting and transporting plants, christmas trees, wreaths, agricultural crops such as potatoes, corn, tobacco, transplants, stone and citrus fruits. *Aquatrols Corp of Am.*

119 Furlong®
For the agriculture industry. *DuPont UK.*

120 Fusion®
For the agriculture industry . *DuPont UK.*

121 Genie®
For the agriculture industry. *DuPont UK.*

122 Granosan®
For the agriculture industry. *DuPont UK.*

123 Grid®
For the agriculture industry. *DuPont UK.*

124 Harvestra®
For the agriculture industry. *DuPont UK.*

125 Horizon®
For the agriculture industry. *DuPont UK.*

126 Hortag Tecnacarb Dust
Carbendazim and tecnazene; protectant fungicide and sprout suppressant for stored potatoes. *Avon Packers Ltd. See* carbendazim, tecnazene.

127 House Plant Leaf Shine
Aerosol containing leafshine material; house plant spray. *Vitax Ltd.*

128 Humber
Granular or powder organic based fertilizers comprising organic base with N, P and K; steady release, lower nitrogen fertilizer for agricultural crops and grasslands. *Humber Fertilizers plc.*

129 Hydan®
For the agriculture industry. *DuPont UK.*

130 Initial®
For the agriculture industry. *DuPont UK.*

131 Jubilee®
For the agriculture industry. *Du Pont UK.*

132 Justice®
For the agriculture industry. *DuPont UK.*

133 K-Cop
Copper-ammonium complex; fungicide. *Griffin.*

134 Leefex
A hop defoliant. *Plant Protection.*

135 Lethane
Insecticide concentrates supplied in petroleum

distillate; used in industrial insecticide sprays and mosquito larvicides. *Rohm & Haas UK.*

136　　Lignasan®
For the agriculture industry. *DuPont UK.*

137　　Liquid Copper Fungicide
Fungicide. *Murphy Chemical Co Ltd.*

138　　Londax®
For the agriculture industry. *DuPont UK.*

139　　Lorate®
For the agriculture industry. *DuPont UK.*

140　　Lupromag®
Magnesium propionate.
Mineralized single feedstuff; animal feed. *BASF AG.*

141　　Lyric®, Lyril®
For the agriculture industry. *DuPont UK.*

142　　Matrix®
For the agriculture industry. *DuPont UK.*

143　　Medley®
For the agriculture industry. *DuPont UK.*

144　　Medo
215-293-2
Oil-based soap containing cresylic acid; pruning compound, canker cure for garden trees. *Vitax Ltd.*

145　　Merfusan
Horticultural product. *May & Baker Ltd.*

146　　Midas®
For agriculture. *DuPont UK.*

147　　Monterey Signal
A blue colored dye to put into the spray tank to let you know where you are spraying; avoid skips and overlaps; breaks down in sunlight. *Lawn & Garden Products Inc.*

148　　Nustar®
For agriculture. *DuPont UK.*

149　　Option®
For the agriculture industry. *DuPont UK.*

150　　Orthomatic
Lawn sprayer. *Monsanto (Solaris).*

151　　Packman
Closed fill system pack opener. *Schering Agrochemicals Ltd.*

152　　Preditec®
For agriculture industry. *DuPont UK.*

153　　Presite®
For agriculture industry. *DuPont UK.*

154　　Propcorn
Chemical products for the treatment of corn. *BP Chemicals Ltd.*

155　　Prospect®
For the agriculture industry. *DuPont UK.*

156　　Quantum®
For the agriculture industry. *DuPont UK.*

157　　Refine®
For the agriculture industry. *DuPont UK.*

158　　Sanction®
For the agriculture industry. *DuPont UK.*

159　　Sirdate®
For the agriculture industry. *DuPont UK.*

160　　Status®
For agricultural applications. *DuPont UK.*

161　　Sympathy®
For the agricultural industry. *DuPont UK.*

162　　Symphony®
For the agricultural industry. *DuPont UK.*

163　　Up-Start
Plant starter. *Monsanto (Solaris).*

164　　Valinate®
For the agriculture industry. *DuPont UK.*

165　　Vi-Grow
Chemical products used for physical conditioning of soil. *Coutaulds plc.*

166　　Vitesse®
For agricultural applications. *DuPont UK.*

Animal Feeds

167　　Ameribond
8061-52-7
Calcium lignosulfonate.
Darran 404; lignosulfonic acid, calcium salt; lime fractionated, spent pulping liquor, precipitate; OSDS-100. Modified calcium lignosulfonates; pelleting aids for animal feeds. *Borregaard Ligno.*

168 Choline Chloride
67-48-1   2261   200-655-4
$C_5H_{14}ClNO$
2-Hydroxy-N,N,N-trimethylethanaminium chloride.
choline hydrochloride; 2-(hydroxyethyl)trimethyl-ammonium chloride; Biocolina; Hepacholine; Lipotril. Animal feed additive. Freely soluble in H_2O, EtOH; LD_{50} (rat orl) = 6.64 g/kg. *Am. Biorganics; Mitsubishi Gas Chemical; Penta Mfg.; Tanabe USA; UCB SA.*

169 Cupric Carbonate
12069-69-1   2697   235-113-6
$CuCO_3.CuH_2O_2$
Basic cupric carbonate.
malachite; [μ-[carbonato(2-)-O:O']]dihydroxydi-copper; (carbonato(2-))dihydroxydicopper; copper carbonate hydroxide; cupric subcarbonate; copper(II)hydroxide carbonate; basic cupric carbonate; Bremen blue; Bremen green; Carbonato(2-(O:O'))dihydroxydicopper; basic copper carbonate ($Cu_2(OH)_2CO_3$); copper carbonate hydroxide; cupric subcarbonate; copper(II)hydroxide carbonate; Basic cupric carbonate; Bremen blue; Bremen green; Basic copper carbonate (Cu2(OH)2CO3); Copper carbonate, various grades. Used in pigments, pyrotechnics, insecticides, copper salts. An astringent in pomades, antidote for phosphorus poisoning, smut preventive, fungicide for seed treatment, feed additive. mp = 200° (dec); d = 4.000; LD_{50} (rat orl) = 1350 mg/kg. *Am. Chemet; Boliden Intertrade; Nihon Kagaku Sangyo.*

170 MHA
583-91-5   6054   209-523-0
$C_5H_{10}O_3S$
2-Hydroxy-4-(methylthio)-butanoic acid.
Methionine hydroxy analog; 2-hydroxy-4-(methylthio)butyric acid; Mha acid; DL-2-hydroxy-4-(methylthio)butyric acid (68-72% in H_2O). Calcium feed supplement for poultry and other animal feeds. *Monsanto Co.*

171 Milk Sugar
63-42-3   5356   200-559-2
$C_{12}H_{22}O_{11}$
D-(+)-Lactose.
Lactose; Milk sugar; 4-O-β-D-galactopyranosyl-D-glucose; β-lactose; β-D-Lactose; Lactose; Lac; lactin; 4-(β-D-galactosido)-D-glucose; lactobiose; saccharum lactin; (+)-β-D-lactose. Used to supplement animal feeds. *See lactose.*

172 Tronacarb Sodium Bicarbonate
144-55-8   8726   205-633-8
$CHNaO_3$
Sodium bicarbonate.
Baking soda; Sodium acid carbonate; Sodium Hydrogen Carbonate; Bicarbonate of soda; Carbonic acid monosodium salt; carbonic acid sodium salt (1:1); col-evac; jusonin; monosodium hydrogen carbonate; monosodium carbonate; meylon; NEUT; soda mint; sodium hydrocarbonate; soludal. White granular solid, industrial and animal feed grades. *Kerr-McGee Chemical Corp. See* sodium bicarbonate.

173 Vita Zinc
1314-13-2   10279   215-222-5
OZn
Zinc oxide.
Low purity zinc oxide; Zinc white; Flowers of zinc; C.I. pigment white 4; Zinc oxide, fume. Animal feed supplement. *Manchem Ltd.*

Carbohydrate Feed Additives

174 Glutalys®
Maize gluten; for animal feed. *Roquette (UK) Ltd.*

175 Molascuit
A cattle food. It is the fine fiber of the sugar cane or begasse, with cane molasses absorbed by it.

176 Molassine Meal
A mixture of molasses and peat moss; cattle feed.

177 Nutranel
Protein, fat, carbohydrate, vitamins, minerals, trace elements; liquid feed. *Roussel Laboratories Ltd.*

178 Pomace
The residue from the extraction of apple juice in cider manufacture; a cattle food.

Mineral-Containing Feed Additives

179 Copper Green
malachite. A term applied to the mineral malachite. Used in pigments, pyrotechnics, insecticides, copper salts. An astringent in pomades, antidote for phosphorus poisoning, smut preventive, fungicide for seed treatment, feed additive. *See* cupric carbonate.

180 Mineral Yeast
Torula, a yeast-like organism; used for fodder production.

Oil-Containing Feed Additives

181 Enervite®
Fish oils, fish liver oils, cod liver oil B. vet C; dietary supplement for animals; conditioning oil for animals. *Seven Seas Ltd.*

Phosphorus Sources

182   Magnaphoscal®
Multiple phosphate (sodium, calcium, magnesium phosphate) granulated; for mineral feeds and mixed feeds. *Bayer AG.*

Protein Sources

183   Alburex
Vegetable proteins; used in animal feedstuff. *Roquette (UK) Ltd.*

184   Pruteen
Single-cell protein used as a feed additive. *ICI Chem & Polymers Ltd.*

Vitamin Sources

185   Ossivite
A proprietary preparation of bonemeal and vitamins A and D. *Wyeth Laboratories.*

186   Pecutrin®
Vitaminized mineral salt mixture; for individual dosing and as a feed additive for all animals kept for use. *Bayer AG.*

Miscellaneous Feed Additives

187   Albutannin
Protan.
Albumen tannate, used in animal feed.

188   Granstock
Animal feed additive. *ICI Chem & Polymers Ltd.*

189   Nosifeed 40
Feed additive (growth promoter) for swine and poultry. *Mitsubishi Kasei.*

190   Nosiheptide
Feed additive (growth promoter) for swine and poultry. *Mitsubishi Kasei.*

191   Nutralys
Animal feedstuff. *Roquette (UK) Ltd.*

192   Saporin®
Feed industry additive. *BASF AG.*

Fertilizers

193   Ammonium Nitrate
6484-52-2   567   229-347-8
$H_4N_2O_3$
Ammonium nitrate.

Ansax; Hero-Prills; Nitram; Old Plantation; Herco-Prills; Nitrammite. Fertilizer prilled ammonium nitrate. Used in explosives, pyrotechnics, herbicides/insecticides, manufacture of nitrous oxide, absorbent for nitrogen oxides, ingredient of freezing mixtures, oxidizer in solid rocket propellants, nutrient for antibiotics and yeast, catalyst. Ammonium nitrate prills containing a minimum of 33.5% nitrogen; used alone or for bulk blending of mixed fertilizers. *L & K Fertilizers Ltd.; Hercules; ICI Chem & Polymers Ltd.; Columbia Nitrogen Corporation; Faith, Keyes and Clark; Air Prods; Chevron; La Roche Ind; Norsk Hydro A/S; Unocal.*

194   Ammonium Phosphate Monobasic
7722-76-1   577   231-764-5
H_6NO_4P
Monoammonium phosphate.
Phosal; Ammonium biphosphate; Ammonium Dihydrogen Phosphate; Ammonium phosphate; ADP; Phosphoric acid, monoammonium salt; Monoammonium phosphate. Used for fertilizers, other chemicals. *Rhône-Poulenc NV; Scottish Agricultural Industries plc.*

195   Ammonium Sulfate
7783-20-2   590   231-984-1
$H_8N_2O_4S$
Sulfuric acid diammonium.
Diammonium sulfate; sulfuric acid, diammonium salt; ammonium sulfate (2:1); ammonium sulfate. Fertilizers, water treatment, fermentation, fireproofing compositions, viscose rayon, tanning, food additive Orthorhombic crystals or white granules; d = 1.7690; mp = 280° (dec); soluble in H_2O (77 g/100 ml); insoluble in alcohol, acetone; pH of 0.1 aqueous solution = 5.5; LD_{50} (rat orl) = 2840 mg/kg. *Acurate Chem & Scientific; Aldrich; AlliedSignal; BASF; DSM NV; General Chem; Heico; Nissan Chem Ind; Schaefer Salt & Chem; Showa Denko.*

196   Ammonium Thiocyanate
1762-95-4   597   217-175-6
CH_4N_2S
Ammonium rhodanide.
Thiocyanic acid, ammonium salt; ammonium sulfocyanate; ammonium sulfocyanide; ATC; ammonium rhodantate; ammonium rhodonide; ammonium rhodanide; Trans-aid. Analytical chemistry; thiourea; fertilizers; photography; in liquid rocket propellants; fabric dyeing; zinc coating; weed killer, defoliant; adhesives; curing resins; pickling iron and steel; electroplating; polymerization catalyst; metals separation. mp = 149°; d = 1.3050; soluble in H_2O (163 g/100 ml); LD_{50} (rat orl) = 750 mg/kg. *Carbo-Tech GmbH; Degussa; Witco/Argus.*

197 Calcium Dihydrogen Phosphate

7758-23-8   1740   231-837-1

CaH_4O8P_2

Monocalcium phosphate anhydrous.

V-90®; calcium phosphate, monobasic; acid calcium phosphate; calcium biphosphate; monocalcium orthophospate; monocalcium phosphate; primary calcium phosphate; Calcium Bis(di-hydrogenphosphate)Monobasic; Phosphoric acid, calcium salt (2:1); calcium bis(dihydrogen-orthophosphate); Calcium dihydro-gen phosphate; calcium dihydrogen phosphate, monohydrate. Chiefly used in fertilizers. Also leavening agent for baking, cereal. Dec 200°; d_4^{18}= 2.220. *Rhône-Poulenc Food Ingreds.*

198 Calcium Orthophosphate

7758-87-4   1741   231-840-8

$Ca_3O_8P_2$

Calcium phosphate tribasic.

Tricalcium Phosphate; Bone Flour; Calcium Phosphate; Phosphoric acid, calcium salt (2:3); Bone phosphate; Calcium phosphate (3:2); Calcium tertiary phosphate; Phosphoric acid, calcium(2+) salt (2:3); Ephos; Tertiary calcium phosphate; Tribasic calcium phosphate (Ca3(PO4)2); Tricalcium diphosphate; Tricalcium orthophosphate. A basic phosphate, containing 60-65% tricalcium phosphate.

199 Copper(II) Sulfate Pentahydrate

7758-99-8   2722   231-847-6

$CuSO_4.H_2O$

Cupric sulfate pentahydrate.

cupric sulphate, pentahydrate; copper(II) sulphate pentahydrate; bluestone; blue vitriol; Roman vitriol; Salzburg vitriol; Kocide® Copper Sulfate Pentahydrate Crystals. Fungicide to control plant diseases, in fertilizers to correct copper deficiencies in soils. Becomes anhydrous by 250°; $d_4^{15.6}$ = 2.286. *Griffin.*

200 Kainite

1318-72-5

$MgK_2S_2O_8.Cl+62Mg.6H_2O$

Magnesium sulfate/magnesium chloride.

Salt found in the Stassfurt deposits; consists mainly of potassium magnesium sulfate and magnesium chloride; consists of a mixture of kainite and rock salt; used in chemicals and fertilizers.

201 Lime Nitrate

10124-37-5   1729   233-332-1

$CaN_2O_6.4H_2O$

Calcium nitrate.

Lime nitrate; Nitric Acid, Calcium Salt; Lime Saltpeter; Norwegian Saltpeter; Nitrocalcite; Calcium Nitrate-15N2. (Lime saltpeter). Fertilizer.

202 Monetite

7757-93-9   1739   231-826-1

$CaHO_4P$

Calcium phosphate dibasic.

Calcium monohydrogen phosphate; Secondary calcium; Calcium Phosphate Dihydrate; Phosphoric acid, calcium salt (1:1); calcium hydrogenorthophosphate; Calcium Hydrogen Phosphate. A calcium phosphate, found in guano.

203 Sodium Borate

1330-43-4   8733   215-540-4

$B_4Na_2O_7$

Boric acid tetrasodium salt.

FB 48; Sodium Borate; Sodium Tetraborate; Tetraborate; Boron sodium oxide (B4Na2O7); Borax, fused; Sodium tetraborate (Na2B4O7); disodium tetraborate. Sodium borate. A fertilizer. *U.S. Borax & Chem.*

204 Sodium Chloride

7647-14-5   8742   231-598-3

$ClNa$

Hydrochloric acid sodium salt.

Betrox. Sodium chloride fertilizer. *ICI Chem & Polymers Ltd.*

205 Superphosphate

1314-56-3   7512   215-236-1

$O_{10}P_4$

Phosphorus pentoxide.

Phosphorous pentaoxide; Phosphoric; phosphoric anhydride; phosphoric pentoxide; Phosphorus(V) oxide; Phosphorus oxide; diphosphorus pentaoxide; Diphosphorus pentoxide. Soluble powder containing 18% phosphorus pentoxide; phosphate fertilizer for use throughout the garden. *Vitax Ltd.*

206 Urea

57-13-6   10005   200-315-5

CH_4N_2O

Carbonyl diamine.

Superprill; Bubber Shet; Carbamide; Carbonyl diamine; Carbamimidic acid; Isourea; aquadrate; ureaphil; ureophil; Aquacare/HP; Nutriplus; Urecare; Urederm. Urea (prilled); fertilizer. Prilled urea containing minimum of 46% available nitrogen; used as a fertilizer to supply nitrogen to the crops for better yield per acre. *Columbia Nitrogen Corporation; Dawood Hercules Chemicals Ltd.*

Metal-Containing Fertilizers

207 Grasshopper

Compound fertilizer 8:1.5:1.5 plus 2,4-D, dicamba and ferrous sulfate; lawn fertilizer. *ICI Garden Products.*

208   **Greenkeeper Mosskiller**
Turf fertilizer with iron sulfate. *Fisons plc, Horticultural Div.*

209   **Kwlk-Green**
Nitrogen, sulfur, iron and zinc; used on turf, shrubs, trees, and potted plants to promote deep rich green foliage. *Lawn & Garden Products Inc.*

210   **Libspray®**
A range of complete foliar fertilizers containing chelated trace elements in powder and liquid form to supplement soil applied fertilizers. *Allied Colloids Ltd.*

211   **Monterey 30% Iron**
Iron sulfate.
30% Iron; ferric sulfate. A granular material used to correct iron deficiency; used on turf, flower beds, vegetables, etc. *Lawn & Garden Products Inc.*

212   **Mosskil**
Lawn fertilizer with iron sulfate. *Fisons plc, Horticultural Div.* See ferric sulfate.

213   **Multigreen® II**
Chelated micronutrients; water-soluble organic blend of metal chelates of iron, zinc, copper and manganese, to improve root growth, color and stress tolerance. *Regal Chemical Company.*

214   **SHL Lawn Sand Plus**
Dichlorophen + ferrous sulfate; moss killer/fertilizer mixture for turf. *Sinclair Horticulture & Leisure Ltd.* See dichlorophen, ferrous sulfate.

215   **Soy-che**
Zinc, iron, copper, manganese, sulfur; chelated micronutrient for soybeans. *Draxel Chemical Company.*

Nitrogen Sources

216   **Agramm**
Nitrogenous fertilizers. *ICI Chem & Polymers Ltd.*

217   **Agriben**
Manure composter for processing liquid and solid manure for agriculture. *Süd-Chemie AG.*

218   **Basfoliar® 34**
Liquid nitrogenous foliar fertilizer with magnesium and micronutrients; for agricultural crops, vines, fruit, hops, and field vegetables. *BASF AG.*

219   **Calurea**
Urea/calcium nitrate.

A nitrogenous fertilizer. It is a mixture of urea and calcium nitrate containing 34% nitrogen.

220   **Ibdu®**
Slow release nitrogen fertilizer. *Mitsubishi Kasei.*

221   **Kalammon**
A fertilizer containing 17% nitrogen and 30% calcium carbonate. *See* calcium carbonate.

222   **Kaliammon saltpeter**
7815
A potassium ammonium nitrate prepared by mixing equivalent molecular proportions of solid potassium chloride and ammonium nitrate in the presence of a little water; a fertilizer. *See* potassium chloride; ammonium nitrate.

223   **Kalluzoto**
A fertilizer containing nitrogen, potassium, and organic matter manufactured from residual molasses.

224   **Kaynitro**
Concentrated nitrogen/potash fertilizer. *ICI Chem & Polymers Ltd.*

225   **Leuna Saltpeter**
A double salt of ammonium sulfate and nitrate; a fertilizer similar to Chilean nitrate in its action.

226   **Longlife Turf Foods**
Fertilizers containing long-lasting nitrogen for sports grounds and parks. *Scottish Agricultural Industries plc.*

227   **Nitraprill**
34.5% Nitrogen prilled fertilizer. *Kemira Ince Ltd.*

228   **Nitro-26**
A nitrogeneous fertilizer. *Fisons plc.*

229   **nitrolim**
Commercial nitrolim contains 57-63% calcium cyanamide, 20% lime, 14% graphite, and 7-8% silica, iron oxide, and alumina; a fertilizer.

230   **Nitrophos® 20-20-0**
Complex fertilizer with 20% nitrogen, 20% phosphate; for all agricultural and horticultural crops. *BASF AG.*

231   **Nitrophoska® 10-15-20**
P_2O_4, 20% K_2O
Complex fertilizer with 10% N, 15% for agricultural and horticultural crops with low basal demand for nitrogen. *BASF AG.*

232   N-Serve
A line of nitrogen stabilizers based primarily on nitropyrin. *Dow UK.*

233   Nufol
Nitrogenous fertilizers. *ICI Chem & Polymers Ltd.*

234   Nuram
Nitrogenous fertilizers. *ICI Chem & Polymers Ltd.*

235   Potazote
A French fertilizer containing 14% nitrogen, as ammonium chloride, and 20% potassium oxide as potassium chloride.

236   Surophosphate
Dasag. A German fertilizer made from sewage, other waste material, and peat.

237   Uramon
A proprietary trade name for a fertilizer containing 43% nitrogen in the form of urea or similar compounds.

Phosphorus Sources

238   African Phosphates
Mineral phosphates found in Tunis and Algeria. They contain from 55-65% calcium phosphate. Others found at Safaga and Kosseir contain 60-70% calcium phosphate. Used as fertilizers.

239   Basfoliar® 6-12-6
Liquid foliar fertilizer with phosphate, zinc copper, and other micronutrients; for Indian corn and other crops with high phosphate requirements. *BASF AG.*

240   Bor-Nitrophoska® 13-13-21+0.1B
Complex fertilizer with 13% nitrogen, 13% phosphate, 21% potash, and 0.1% boron; for all agricultural crops requiring boron and horticultural crops which are not sensitive to chloride. BASF AG.

241   Cederan® P 23
Phosphate single fertilizer (23% phosphate); for all crops and soil types. *BASF AG.*

242   Conifer and Shrub Fertiliser
Powdered fertilizer containing NPK 10:7.5:10 plus 1.85 Mg and trace elements; all-purpose base and top dressing fertilizer. Vitax Ltd.

243   Eclipse
Granular or powder organic based fertilizers; a steady release, lower nitrogen fertilizer for horticultural crops, parks and gardens. *Humber Fertilizers plc.*

244   Enmag
Magnesium ammonium phosphate fertilizer. *Scottish Agricultural Industries plc.*

245   Folex-P
Foliar phosphate-based fertilizer. *Omex Agriculture Ltd.*

246   GoodLife
A powder organic based fertilizer comprising a composted organic base and chemical N, P & K to form the analysis; four types sold: all-purpose Fertilizer, flower garden fertilizer, vegetable fertilizer and lawn weed and feed fertilizer. *Humber Fertilizers plc.*

247   Green Up Autumn Liquid Lawn Feed
Liquid concentrate containing NPK 3:6:6 and 1% Fe; autumn feed for lawn areas. *Vitax Ltd.*

248   Green Up Feed and Weed Plus Moss Killer
Dry powder containing NPK 8:4:4 plus 2,4-D, mecoprop and ferrous sulfate; combined fertilizer, weed and moss killer. *Vitax Ltd.*

249   Green Up Lawn Feed and Weed
Liquid concentrate containing NPK 14.5:3:3 and 2,4-D and dicamba; combined feed and weed for turf. *Vitax Ltd.*

250   Green Up Lawn Feed 'n Weed Plus Moss Killer
Dichlorophen + mecoprop + dichlorprop + dicamba + benazolin.
Liquid concentrate containing NK 11:4 plus dichlorophen, mecoprop, dichlorprop, dicamba and benazolin; combined feed, weed and moss killer. *Vitax Ltd.* See dichlorophen, mecoprop, dichlorprop, dicamba, benazolin.

251   Green Up Liquid Lawn Feed
Liquid concentrate containing NPK 17:3.5:3.5; liquid fertilizer for turf. *Vitax Ltd.*

252   GRO-HY
Nitrogen, phosphorus, potash plus trace elements as a slow-release fertilizer tablet; fertilizer for trees, shrubs and bushes. *Envhy Ltd.*

253   Growmore
Granular fertilizer containing NPK 7:7:7; general-purpose fertilizer. *Vitax Ltd.*

254   Guano
Bird Manure.
Consists of deposits of excrement and skeletons of birds and animals; a fertilizer rich in phosphorus and nitrogen.

255   House Plant Liquid Feed
Liquid concentrate containing NPK 5:2:2; house plant feed. *Vitax Ltd.*

256   Humber
Granular or powder organic based fertilizers comprising organic base and chemical N, P and K to form the analysis; a steady release, lower nitrogen fertilizer for agricultural crops and grasslands. *Humber Fertilizers plc.*

257   Leunaphos
A mixture of phosphate, nitrate, and sulfate of ammonia. Fertilizer.

258   Liquid Feed for Hanging Baskets
Liquid concentrate containing NPK 4:2:6 plus trace elements; liquid fertilizer for hanging baskets, tubs, planters and window boxes. *Vitax Ltd.*

259   Liquid Growmore
Liquid concentrate containing NPK 7:7:7: general purpose liquid fertilizer. Vitax Ltd.

260   Liquid Q4 Borders and Beds Fertiliser
Liquid concentrate containing NPK 5.3:5.3:10 plus trace elements; border and bedding plant fertilizer. *Vitax Ltd.*

261   Liquid Tomato Feed
Liquid concentrate containing NPK 4.5:4.5:9 plus Mg; fertilizer. *Vitax Ltd.*

262   Monterey Foliar Nutrient 11-4-6
Used as a foliar nutrient, providing nitrogen, phosphoric acid, potash and chelated zinc, iron and manganese. For use on turf, ornamentals, trees, vegetables and house plants. *Lawn & Garden Products Inc.*

263   Multi Base
Powdered compost additive containing NPK 2.6:2.2:23 plus 3% Mg and trace elements; compost additive. *Vitax Ltd.*

264   Neutral Phosphate
A fertilizer prepared by digesting mineral phosphate, bonemeal, or a mixture of both, with small amounts of sulfuric acid. This renders the P_2O_5 more available. The product contains 20-25% P_2O_5, and is neutral.

265   Nitrophosphate
A fertilizer sometimes wrongly called ammonium superphosphate. It is prepared by mixing calcium superphosphate with ammonium sulfate giving mixtures containing ammonium phosphate and calcium sulfate.

266   Pestilizer®
Phosphate esters.
Compatibilizers for liquid fertilizers. *Stepan.*

267   Phosphazote
Mixture of superphosphate and urea containing 4-11% nitrogen and 10-14% P_2O_5; Fertilizer.

268   Polyfeed
Water soluble, chlorine free N-P-K fertilizer, with chelated micro-nutrients; for direct soil application, via irrigation system or foliar spray. *Haifa Chemicals Ltd.*

269   Potassic Superphosphate
A fertilizer made by combining calcium superphosphate with potash salts.

270   Potting Base
Powdered compost additive containing NPK 3.6:2.2:2.9 plus 3% Mg and trace elements; compost additive. *Vitax Ltd.*

271   Rhenania phosphate
Vesta phosphate.
Prepared by sintering together in a furnace at 1200-1300 C, a mixture of raw phosphate, limestone, and alkali silicate. Resulting product approximates to the formula $Ca_2KNa(PO_4)_2$.

272   Saltpeter Superphosphate
A fertilizer made by mixing niter with calcium superphosphate.

273   Scotphos
Fertilizers containing phosphate. Scottish Agricultural Industries plc.

274   Silico-superphosphate
Preparation made by mixing superphosphate with kieselguhr or precipitated silicic acid. It is stated to give better results on medium and light soils.

275   Soluble Plant Feed
Soluble powder containing NPK 19:19:19 plus 0.2% Mg and trace elements; a fertilizer. *Vitax Ltd.*

276   Soluble Rose Feed
Soluble powder containing NPK 16:8:32 plus 0.15% Mg and trace elements; fertilizer. *Vitax Ltd.*

277   Soluble Tomato Feed
Soluble powder containing NK 18:36 and trace elements; a fertilizer. *Vitax Ltd.*

278   Steamed Bone Meal
A fertilizer consisting of crushed bones treated

with superheated steam and benzene to remove fat and glue; contains about 1% nitrogen.

279   Stoffertite
$CaHPO_4 \cdot 5H_2O$
A calcium phosphate, it occurs in guano.

280   Superam
A fertilizer obtained by neutralizing the acids of ordinary super phosphate with ammonia gas.

281   Superphosphate
$CaH_4O_8P_2$
Mineral superphosphate; superphosphate of lime. Consists of mono-calcium phosphate, mixed with calcium sulfate, and contains 25-28% soluble phosphate; used as a fertilizer.

282   Swardsman
25-5-5, compound fertilizer. *Kemira Ince Ltd.*

283   Tennessee Phosphates
Mineral phosphates containing from 60-70% calcium phosphate; used as a fertilizer.

284   Treble Superphosphate
Monocalcium phosphate, containing 48-49% P_2O_5 (41-42% water soluble P_2O_5); fertilizer.

285   Ureaphos
A fertilizer containing phosphate of ammonia and urea.

286   Vitafeed 101
Soluble powder containing NK 26:26 and trace elements; fertilizer. *Vitax Ltd.*

287   Vitafeed 102
Soluble powder containing NK 18:36 and trace elements; fertilizer. *Vitax Ltd.*

288   Vitafeed 103
Soluble powder containing NK 13:43 and trace elements; fertilizer. *Vitax Ltd.*

289   Vitafeed 111
Soluble powder containing NPK 19:19:19 plus 0.2% Mg and trace elements; fertilizer. *Vitax Ltd.*

290   Vitafeed 301
Soluble powder containing NK 36:12 plus 0.15% Mg and trace elements; fertilizer. *Vitax Ltd.*

291   Vitax Q4
Powder fertilizer containing NPK 5.3:7.5:10 plus 1.8% Mg and trace elements; all-purpose base and top dressing fertilizer. *Vitax Ltd.*

292   Wiborg Phosphate
A German fertilizer made by heating mineral phosphate with soda. It consists mainly of a tetraphosphate.

293   Zewaphosphate
A phosphate fertilizer.

Urea Sources

294   Fluf® 10-0-10
Urea-formaldehyde suspension; fertilizer providing nitrogen and potassium to golf course turf. *W A Cleary.*

Miscellaneous Fertilizers

295   Agrosil® LR
Colloidal silicate.
Encourages intensive root development, improves irrigation efficiency, improves soils. *BASF AG.*

296   Agro-Vita
Mineral and plant extracts in a water base containing cytokinin, B-vitamin, morphogenic and porphyrin. Activity to aid in increased plant metabolism and yield; for all agricultural, horticultural and forestry products. *SN Corp/Appropriate Technology Ltd.*

297   Aither's Lawn Sand Plus
Dichlorophen + ferrous sulfate.
A moss killer/fertilizer mixture for turf. *R. Aitken. See* dichlorophen, ferrous sulfate.

298   Albert
Basic slag used for fertilizing purposes. *Canadian Hoechst.*

299   Amdye PH-12
Inorganic salts; replacement for trisodium phosphate in liquid form. *Am. Emulsions.*

300   Autumn Lawn Food
Lawn fertilizer. *Fisons plc, Horticultural Div.*

301   Bayfolan®
Foliar feed containing macro and micro nutrients; for all agricultural and horticultural crops to help recover from the effects of adverse conditions such as drought, low temperatures or waterlogging. *Bayer AG; Bayer plc.*

302   Bilston
A basic slag used for fertilizing purposes. *Fisons plc, Horticultural Div.*

303   Cutonic
Micronutrient foliar sprays. *McKechnie Chemicals Ltd.*

304   Deep Feed
Liquid fertilizer. *Fisons plc, Horticultural Div.*

305   Equalized Guano
Natural guanos, blended or mixed with ammonium salts, to obtain proportions of nitrogen and phosphorus; a fertilizer.

306   Evergreen
Lawn fertilizer combined with selective weedkiller. *Fisons plc, Horticultural Div.*

307   Felspar
$(K_2O \cdot 3SIO_2).2(Al_2O_3 \cdot 3SIO_2)$
Potassium Felspar.
Orthoclase. A potassium aluminum silicate, used in the manufacture of porcelain, as a building material, and as a fertilizer.

308   Fetrilon® Combi
Easily soluble micronutrient mixed fertilizer containing 9% magnesium; for all agricultural crops, vines, fruit, and hops. *BASF AG; BASF plc.*

309   Ficote
Range of coated fertilizers. *Fisons plc, Horticultural Div.*

310   FL7P
Liquid fertilizer. *Fisons plc, Horticultural Div.*

311   Folia-Feed
Foliar nutrient. *DowElanco Ltd.*

312   Foliar 36 Extra
Foliar feed . *BASF plc.*

313   Foliar Nitrophoska
Foliar feed. *BASF plc.*

314   GH5
Granular fertilizer. *Fisons plc, Horticultural Div.*

315   Glasgro
Range of granular fertilizers. *Fisons plc, Horticultural Div.*

316   Green Magic
Various granular fertilizer blends for lawns, gardens and flowers. *Horn's Crop Service Center.*

317   Greenkeeper
Range of turf fertilizers. *Fisons plc, Horticultural Div.*

318   Gypsum-F
Source of calcium and sulfur in fertilizers for use on golf courses, ball parks, cemeteries, nurseries, industrial lawns, home lawns, and other turf applications; increases permeability of soils to lower sodium content. *W A Cleary.*

319   Horn O'Plenty
Various granular fertilizer blends; farm and garden fertilizers. *Horn's Crop Service Center.*

320   Hortus
Fertilizers. *Scottish Agricultural Industries plc.*

321   Houseplant Long Lasting Feed
Coated fertilizer. *Fisons plc, Horticultural Div.*

322   Insect Spray for House Plants
Contact fertilizer aerosol. *Fisons plc, Horticultural Div.*

323   Kericompost
Compost for houseplants. *ICI Garden Products.*

324   Kerigrow
Compound fertilizer 6:4:4; liquid fertilizer for houseplants. *ICI Garden Products.*

325   Keriguards
Pellets containing fertilizer and insecticide in combination. *ICI Chem & Polymers Ltd.*

326   Kerispikes
Compound fertilizer 6:4:4; food spikes for houseplants. *ICI Garden Products.*

327   Keristicks
Capilliary sticks for houseplants. *ICI Chem & Polymers Ltd.*

328   Lawn Food
Lawn fertilizer. *Fisons plc, Horticultural Div.*

329   Lawnsman
Range of lawn aids for garden use such as fertilizers, weedkillers, and a spreader. *ICI Garden Products.*

330   Lawnsman Spring Feed
Spring/summer lawn food. *ICI Chem & Polymers Ltd.*

331   Lawnsman Weed and Feed
Fertilizer combined with selective weedkiller. *ICI Chem & Polymers Ltd.*

332   Lawnsman Winterizer
Autumn lawn feed. *ICI Chem & Polymers Ltd.*

333   Liquinure
Liquid fertilizer. *Fisons plc, Horticultural Div.*

334   Magspa
Fertilizer additive. ICI Chem & Polymers Ltd.

335   Mantrilon®
Liquid manganese fertilizer; foliar fertilizer to prevent and cure deficiency of manganese in all agricultural crops, vines and fruit. *BASF AG.*

336   Mantrllon® FL
Foliar feed. *BASF plc.*

337   MG2/MG4
Range of granular fertilizers. *Fisons plc, Horticultural Div.*

338   Monterey Bloom Popper
A 8-32-7 liquid fertilizer formulated with humic acid for better foliar uptake; high phosphate levels help in the blooming and fruit products of a plant and the humic acid helps in the uptake of that nutrient. *Lawn & Garden Products Inc.*

339   Monterey Iron Chelate 10%
Used to correct Iron deficiencies, either as a soil drench or as a foliar spray; may be used on ornamentals, vegetables, fruit trees, etc. *Lawn & Garden Products Inc.*

340   Monterey Perc-O-Late Plus
Surfactants plus nitrogen, zinc, iron and manganese; used for water penetration and fertilization of turf, flower beds, potted plants, etc. *Lawn & Garden Products Inc.*

341   Monterey Stimulator 12
Humic acid derived from completely organic sources; aids the plant in the uptake of nutrients from the soil or through foliar application; helps plants utilize the fertilizer you give them. *Lawn & Garden Products Inc.*

342   Nettolin
Humus complex-fertilizer on peat basis for the culture of wine, hops, fruit and vegetables, for flowers and lawns. *Sûd-Chemie AG.*

343   NFT Fertilizer
Soluble fertilizer. *Fisons plc, Horticultural Div.*

344   Nitracc
Fertilizers. *ICI Chem & Polymers Ltd.*

345   Nitrammomkalk
A mixture of ammonium and calcium nitrates in a granular form of Norwegian manufacture; used as a fertilizer.

346   Nitrapo
A product obtained from crude caliche by crystallization. It contains about 66% sodium nitrate, 29% potassium nitrate, and a little sodium chloride; used as a fertilizer.

347   Nitro-chalk
A proprietary fertilizer consisting of an intimate mixture of chalk and ammonium nitrate in the form of a fine powder. *ICI Chem & Polymers Ltd.*

348   Pioneer
12-20-20 Compound fertilizer. *Kemira Ince Ltd.*

349   Plus
Range of fertilizers for garden use. *ICI Chem.*

350   Polypel
Fertilizer. *Monsanto (Solaris).*

351   Proleaf
Horticultural foliar feed fertilizer. *Schering Agrochemicals Ltd.*

352   Promoloid
5727
A fertilizer containing collodial magnesium silicate; a Japanese product.

353   PS3/PS4/PS5
Range of granular fertilizers. *Fisons plc, Horticultural Div.*

354   Ra-Pid-Gro
Plant food. *Monsanto (Solaris).*

355   Rose Food
Granular fertilizer. *Fisons plc, Horticultural Div.*

356   Rose Plus
Granular fertilizer containing magnesium. *ICI Garden Products.*

357   SeaGarden
Soluble nutrients for algae, particularly in marine aquariums; aquarium water supplement. *Aquarium Systems Inc.*

358   Siapton
Liquid organic foliar feed. *ICI Chem.*

359   Silvital
Fertilizer for vitalizing damaged forest. *Sûd-Chemie AG.*

360   Solinure
Range of soluble fertilizers. *Fisons plc, Horticultural Div.*

361 Solufeed
Soluble fertilizer. *ICI Chem & Polymers Ltd.*

362 Stimufol
Soluble fertilizer. *ICI Chem & Polymers Ltd.*

363 Super Green
Various granular fertilizer blends; fertilizers for lawns, gardens, and flowers. *Horn's Crop Service Center.*

364 Supergreen
Lawn fertilizer. *May & Baker Ltd.*

365 Terrafen®
Special fertilizer incorporated into a gel which operates on the ion exchange principle; for long-term fertilization of all types of houseplants. *Bayer AG.*

366 Thomas Meal
Ground slag obtained from the Thomas process for iron; used as a fertilizer.

367 Tomorite
Liquid fertilizer. *Fisons plc, Horticultural Div.*

368 Triabon® 16-8-12-4
Complex fertilizer for substrates regardless of their pH, and greenhouse crops. BASF AG.

369 Twosward
Fertilizers. *ICI Chem & Polymers Ltd.*

370 UN-28 and UN-32
Fertilizer solutions containing 28% and 32% nitrogen respectively; designed for direct agricultural use as a three-way source of nitrogen. Hercules.

371 Unipel
Pelleted fertilizer. *Monsanto (Solaris).*

372 Velpak
Seaweed based crop stimulant. *Bayer AG.*

373 Verdley
Range of composts and soil conditioners based on peat or bark. *ICI Chem & Polymers Ltd.*

374 Zeotokol
A coarse dolerite (an igneous rock composed essentially of labradorite and anorthite, with augite and sometimes olivine) ground up; and used as a fertilizer.

Fungicides

375 Aaterra WP
2593-15-9   219-991-8
$C_5H_5Cl_3N_2OS$
5-Ethoxy-3-(trichloromethyl)-1,2,4-thiadiazole.
Etridiazole; Banrot; Echlomezol; Dwell; ETCMTB; Ethazol; Ethazole; Ethazole (fungicide).; 5-Ethoxy-3-trichloromethyl-1,2,4-thiadiazol; 5-Ethoxy-3-trichloromethyl-1,2,4-thiadiazole; ETMT; Etridiazol; Etridiazole; Koban;MF-344; Olin Mathieson 2,424; OM 2424. Protective fungicide which is incorporated into soil or compost. *ICI Agrochemicals. See etridiazole.*

376 Acticide PMA 100
62-38-4   7453   200-532-5
$C_8H_8HgO_2$
(Acetato)phenylmercury.
Phenylmercuric acetate; Phenyl mercury acetate powder; PMA; PMAC; PMAS; Ceresan slaked lime; Gallotox; Liquiphene; Mersolite; Nylmerate; Phix; Riogen; Scutl; Tag Fungicide; Tag HL-331; Single Purpose; Merpectogel; Agrosan GN; Ceresol; Single Purpose; PMA 18, 60. Wet-state bactericide/fungicide for emulsion paints, plasters, wood-pulp, etc. Used as a herbicide and fungicide and a seed dressing for cereals and fodder beet. mp = 149°; soluble in H_2O (1.6 mg/ml), more soluble in organic solvents; LD_{50} (rat orl) = 22 mg/kg. *Thor Chemicals (UK) Ltd.; Poythress Laboratories Inc.; ICI Chem & Polymers Ltd.; ICI AgroChemicals; DowElanco Ltd.; Hûls Am.*

377 Afugan
13457-18-6   8146   236-656-1
$C_{14}H_{20}N_3O_5PS$
2-[(Diethoxyphosphinothioyl)oxy]-5-methylpyrazolo[1,5-a]pyrimidine-6-carboxylic acid ethyl ester.
Pyrazophos; Curamil; HOE 2873; Missile. Systemic organophosphorus fungicide. m.p. 38-40°; LD_{50} (rat orl) = 140 mg/kg. *Hoechst UK. See pyrazophos.*

378 Aliette
15845-66-6
$C_2H_7O_3P$
Ethylphosphonic acid.
fosetyl. Fungicide. *May & Baker Ltd. See fosetyl.*

379 Aliette
39148-24-8   4278   254-320-2
$C_6H_{18}AlO_9P_3$
Fosetyl-aluminum.
Fosetyl-Al; Mikal; aluminum tris(ethyl hydrogen phosphonate); Efosite-Al; Epal; EXP 1659; LS

74783; phosethyl-al; RP-32545; Chipco aliette WDG; aluminum triethyl triphosphonate. A systemic phosphonate fungicide for horticultural crops. *Embetec Crop Protection Ltd.* See fosetyl-aluminum.

380   Anilazine
101-05-3   694   202-910-5
$C_9H_5Cl_3N_4$
4,6-Dichloro-N-(2-chlorophenyl)-1,3,5-triazin-2-amine.
Dairene®; Dairin®; Dyrene®; Dyrene; Kemate; Triasyn; Direx; B-622; Bortrysan; Direz; Dyrene 50W Triazine; Zinochlor; Aniyaline; Triasym; Anilazine. Broad spectrum fungicide used for tobacco, potatoes, cereals and ornamentals. Non-systemic foliar fungicide with protective action used to control blights of potatoes and tomatoes and leafspot diseases in many crops. mp = 159-160°; insoluble in H_2O; soluble in toluene, xylene, acetone; LD_{50} (rat orl) >5000 mg/kg. *Bayer AG.*

381   Antifungin
13703-82-7   5693   237-235-5
B_2MgO_4
Magnesium borate.
The trade name for magnesium borate, an antiseptic and fungicide. Occurs in a variety of minerals. Slightly soluble in H_2O.

382   Arsenic Trioxide
1327-53-3   844   215-481-4
As_4O_6
Arsenic (III) oxide.
Arsenic oxide; Arsenous trioxide; arsenous acid; arsenous oxide; arsenic sesquioxide; White Arsenic; Diarsenic Trioxide; Crude Arsenic; Arsenic (white); Arsenious oxide; Arsenic (III) trioxide; Arsenous anhydride; arsenite; arsenolite; arsenous acid anhydride; arsenous oxide anhydride; arsodent; claudelite; claudetite; Arsenic oxide (3); Arsenic oxide (As_2O_3); Arsenic sesquioxide (As_2O_3); Arsenicum album; Diarsonic trioxide; Diarsenic oxide. Pigments, ceramic enamels, aniline colors, decolorizing agent in glass, insecticide, rodenticide, herbicide, sheep and cattle dip, hide preservative, wood preservative, preparation of other arsenic compounds. mp = 315°; bp = 465°; soluble in H_2O, dil HCl, alkali hydroxide or carbonate solns; insoluble in EtOH, $CHCl_3$, Et_2O; LD_{50} (rat orl) = 1.46 mg/kg. *Atomergic Chemetals; Noah Chem.; Outokumpu Oy; Transene.*

383   Ashlade TCNB
117-18-0   204-178-2
$C_6HCl_4NO_2$
1,2,4,5-Tetrachloro-3-nitrobenzene.
Bygran S; Fusarex; Hickstor; Hystor 10; Hytec;

New Hickstor 6; New Hystor; Quad Store; Quad-Keep; Tripart® Arena 6; Tubodust; Tubostore; Nebulin; Tecgran. Granules or dustable powder containing tecnazene; protectant fungicide and potato sprout suppressant. Tecnazene (6-10% w/w); used to control dry rot in both ware and seed potatoes, and sprouting in ware potatoes. Tecnazene in liquid fogging solution; used for controlling sprouting and dry rot in stored potatoes. mp = 99°; bp = 304° (dec); soluble in H_2O (0.44 mg/l), more soluble in organic solvents; LD_{50} (rat orl) = 2047 mg/kg. *Ashlade Formulations Ltd.; Dean Agrochemicals Ltd.; ICI Chem & Polymers Ltd.; Hickson & Welch Ltd.; Agrichem (International) Ltd.; Quadrangle Agrochemicals; Tripart Farm Chemicals Ltd.; Farmers Crop Chemicals Ltd.; Wheatley Chemical Co Ltd.; Atlas Interlates Ltd.* See tecnazene.

384   Atiran
123-88-6   204-659-7
C_3H_7ClHgO
Chloro(2-methoxyethyl)mercury.
Agallol; Aratan; Aretan 6; Baytan®; Ceresan; Ceresan Universal Nazbeize; Chloro(2-methoxyethyl)mercury; Falisan; Gramisan; Higosan; Agallolat; Agalol; Aretan; Atiran; Cekusil Universal C; Ceresan-Universal Nassbeize; MEMC; Merchlorate; (β-methoxyethyl)mercuric Chloride; Methoxyethyl Mercuric Chloride; 2-methoxyethylmercuric Chloride; β-methoxyethylmercury Chloride; 2-methoxyethylmercury Chloride; Sedresan; Tafasan 6W; Tafasan; Triadimenol-fuberidazole. Dry powder containing 25% w/w triadimenol and 3% w/w fuberidazole; a seed treatment for barley, wheat, oats and rye; controls the important seed and certain soil borne diseases including loose smut, covered smut, foot rot, leaf stripe bunt and early attacks of mildew and rhynchosporium. Seed dressing for control of fungal diseases on cereals, rice, cotton and vegetables. mp = 65°; insoluble in H_2O, organic solvents; LD_{50} (rat orl) = 22 mg/kg. *Plant Protection; Bayer AG; Bayer plc.* See methoxyethyl mercury chloride.

385   Bayleton® 5
43121-43-3   9723   256-103-8
$C_{14}H_{16}ClN_3O_2$
Triadimefon.
Bayleton; triamefon; Bonide Bayleton Systemic Fungicide; Green Light Fung-Away Fungicide; SA Systemic Fungicide for Turf & Ornamentals; Amiral; azocene; Bay 6681 f; Bayleton 250 ec; Bay-meb-6447; meb 6447; triadimephon; tidifon; Triadimefon; Rofon; Monterey Bayleton. Wettable systemic fungicide powder containing 5% w/w triadimefon; used to control powdery mildew on apples, hops, raspberries, strawberries and other cane fruits plus American gooseberry mildew on

all varieties of blackcurrants and gooseberries. Also used for the control of diseases in turf and the control of powdery mildew, rusts and blight of ornamental plants. *Bayer plc; Lawn & Garden Products Inc. See* triadimefon.

386   Benomyl

17804-35-2   1073   241-775-7

$C_{14}H_{18}N_4O_3$

[1-[(Butylamino)carbonyl]-1H-benzimidazol-2-yl]carbamic acid methyl ester.

Benlate; F-1991;; Benlate; Tersan 1991; Benlate(R); F1991; Fungicide; Agrocit; Benosan; Du Pont 1991; Fundazol; Benex; BBC; arilate; benlate 50; benlate 50 w; benomyl 50 w; BNM; D 1991; fundasol; fungicide 1991; Uzgen; NS O2; fibenzol; arbortrine. Fungicide for garden use. (Sold in UK on behalf of Du Pont). Insoluble in H_2O, soluble in $CHCl_3$, LD_{50} (rat orl) > 9590 mg/kg. *ICI Chem & Polymers Ltd.*

387   Bitertanol

55179-31-2   1342   259-513-5

$C_{20}H_{23}N_3O_2$

β-[1,1'-Biphenyl)-4-yloxy]-α-(1,1-dimethylethyl)-1H-1,2,4-triazol-1-ethanol.

BAY KWG 0599; Sibutol; Baycor®. Broad spectrum fungicide. Mixture of diastereoisomers. Systemic foliar fungicide used especially to control scab on apples and black spot on roses. mp = 125-129°; slightly soluble in H_2O (5 mg/l), more soluble in organic solvents; LD_{50} (rat orl) > 5000 mg/kg. *Bayer AG.*

388   Butylparaben

94-26-8   1619   202-318-7

$C_{11}H_{14}O_3$

Butyl p-hydroxybenzoate.

4-hydroxybenzoic acid butyl ester; n-butyl p-hydroxybenzoate; Lexgard B; Butoben; Butyl Chemosept; Butyl Parasept; Tegosept B. Antifungal for pharmaceuticals; preservative in foods. mp = 68-69°; soluble in H_2O, organic solvents; LD_{50} (mus orl) = 13.2 g/kg. Inolex; *Nipa Labs; Penta Mfg.*

389   Caddy

10108-64-2   1653   233-296-7

$CdCl_2$

Cadmium turf fungicide.

cadmium chloride; Cadmium dichloride; Caddy; Vi-Cad. *W A Cleary. See* cadmium chloride.

390   Calcium Oxide

1305-78-8   1733   215-138-9

CaO

Lime.

quicklime; calx. Inorganic oxide; refractory, used in sewage treatment, insecticides, fungicides,

manufacture of steel and aluminum; flotation of nonferrous ores; manufacture of glass, paper, Ca salts; in drilling fluids, lubricants; laboratory. mp = 2572°; bp = 2850°; d = 3.32-3.35. *Cerac; GE; Hüls Am.; Mallinckrodt; Pfizer; U.S. Gypsum.*

391   Calirus

15310-01-7   239-352-7

$C_{13}H_{10}INO$

2-Iodo-N-phenylbenzamide.

Benodanil; Apache; BAS 3170F; Benefit; Mascot Clearing. A systemic fungicide with protective and curative action. Used for control of rust disease in various crops. *BASF plc; Rigby Taylor Ltd. See* benodanil.

392   Calixin®

24602-86-6   9793   246-347-3

$C_{19}H_{39}NO$

2,6-Dimethyl-4-tridecylmorpholine.

Tridecyl-2,6-dimethylmorpholine; Tridemorph. Systemic fungicide for control of powdery mildew in cereals, vegetables, etc. *BASF AG; BASF plc. See* tridemorph.

393   Calomel

10112-91-1   5951   233-307-5

Cl_2Hg_2

Mercurous chloride.

Dimercury dichloride; Mercurous chloride. Fungicide. Used for control of clubroot in brassicas and white rot in onions. *Hortichem Ltd. See* mercurous chloride.

394   Campbell's Nabam Soil Fungicide

142-59-6   6426   205-547-0

$C_4H_6N_2Na_2S_4$

Disodium ethylenebisdithiocarbamat.

nabam; Dithane; Chem Bam; Spring Bak; Dithane A-46; EBDC, disodium salt; Ethylenebis(dithiocarbamic acid), disodium salt; Nalco D-62C44. Soluble concentrate containing 320 g/l nabam; used for control of root rot in tomatoes and chrysanthemums. *MTM AgroChemicals Ltd. See* nabam.

395   Captafol

2425-06-1   1814   219-363-3

$C_{10}H_9Cl_4NO_2S$

3a,4,7,7a-Tetrahydro-2-[(1,1,2,2-tetrachloroethyl)thio]-1H-isoindole-1,3(2H)-dione.

Sanspor; Difolatan; Merpafol; Morestan; Morestan®; quinomethionate; oxythioquinox; quinoxalines; ENT 25606; Bay 36205; Chinomethionate; Forstan; SS 2074. Fungicide, used with potatoes. Wettable powder containing 25% w/w quinomethionate; for control of red spider mites, including organophosphorus strains, on apples, gooseberries, strawberries, and marrows and American gooseberry mildew (partial control of leafspot) on gooseberries, mp = 160-161°; LD_{50} (rat orl) = 2500-6200 mg/kg. *Makhteshim Chemical Works Ltd.; ICI Chem & Polymers Ltd.; Hortichem Ltd.; Bayer AG.*

396 Carbendazim
10605-21-7   1836   234-232-0
$C_9H_9N_3O_2$
Carbendazim.
1H-Benzimidazol-2-ylcarbamic acid methyl ester; 2-benzimidazolecarbamic acid methyl ester; 2-(methoxycarbonylamino)benzimidazole) methyl 2-benzimidazolecarbamate; carbendazole; BMC; MBC; BCM; Vinclozolin; BAS-3460; BAS-67054; CTR-6669; HOE-17411; Bavistin; Derosal; Battal FL; BAS 3460; BAS 3460F; BAS 67054F; BCM; BMK; Carbendazime; Carbendazol; Carbendazole; Carbendazym; G 665; Kemdazin; Mecarzole; Konker®; Hinge; Mascot Systemic Turf Fungicide; Maxim; Stempor DG; Tripart® Defensor FL; Turfclear. Fungicide with systemic activity and contact effect with protective and curative action. Controls a wide variety of fungal diseases in cerals, fruit and vegetables. Used against Dutch elm disease. mp = 302-307° (dec); pKa = 4.48. *BASF AG; ICI Chem & Polymers Ltd.; ICI Agrochemicals; Pan Britannica Industries Ltd.; Quadrangle Agrochemicals; Rigby Taylor Ltd.; Farmers Crop Chemicals Ltd.; Tripart Farm Chemicals Ltd.; Fisons plc.*

397 Carbendazim
83601-81-4
$C_{13}H_{18}ClN_3O_2$
(5-Butyl-1H-benzimidazol-2-yl) carbamic acid methyl ester monohydrochloride.
Ashlade Mancarb FL; Ashlade M; Delsene® 50 DF; Derosal WDG. Fungicide for cereals. *Ashlade Formulations Ltd.; DuPont UK; Hoechst UK.*

398 Carboxin
5234-68-4   1874   226-031-1
$C_{12}H_{13}NO_2S$
5,6-Dihydro-2-methyl-N-phenyl-1,4-oxathiin-3-carboxamide.
1,4-Oxathiin-3-carboxanilide, 5,6-dihydro-2-methyl; Carbathiin; Carboxine; D 735; DCMO; DMOC; F 735; Vitavax; Vitavax 100; Vitavax 735D; Vitavax 75W; 1,4-oxathiin, 2,3-dihydro-5-carboxanilido-6-methyl-; Kemikar; Kisvax; Oxatin. Systemic fungicide; used for seed treatment and control of smuts, bunts and seedling diseases. mp = 93-95°; soluble in H_2O (175 mg/l, 25°); more soluble in organic solvents; LD_{50} (rat orl) = 3820 mg/kg. *Uniroyal; Kemira; Jin Hung; Diachem.*

399 Chloroneb 65W Fungicide
2675-77-6   220-222-3
$C_8H_8Cl_2O_2$
1,4-Dichloro-2,5-dimethoxybenzene.

Demosan; Soil Fungicide 1823; Teremec; Tersan SP. 65% Chloroneb wettable powder; seed treatment to suppress seeding blights, soreshin and pre- and post-emergence damp-off caused by *rhizoctonia solani, pythium spp and sclerotium rolfsii* on cotton, beans, soybeans and sugar beets. mp 133-135°; bp = 268°; soluble in H_2O (8 mg/l), more soluble in organic solvents; LD_{50} (rat orl) > 11000 mg/kg. *Kincaid Enterprises Inc.* See 1,4-dichloro-2,5-dimethoxybenzene.

400 Chlorothalonil
1897-45-6   2219   217-588-1
$C_8Cl_4N_2$
2,4,5,6-Tetrachloroisophthalonitrile.
Bombardier; Chiltern Ole; Contact 75; Forturf; Bravo; Bravo 500; Exotherm; m-TCPN; Sweep; TCIN; Termil; TPN; Daconil; DAC-2787; Daconil 2787; Daconil Turf; Exotherm Termil; Chlorothanonil; Farber; Jupital; Ole; Pillarich; Repulse; Taloberg; Tuffcide; Groutcide 75; Black Leaf Lawn & Garden Fungicide; Bonide; ClortoCaffaro; Clortosip; Dragon Daconil 2787; Ferti-lome; Green Charm Multi-Purpose Fungicide; Green Thumb Lawn & Garden Fungicide; Ortho Multi-Purpose Fungicide Daconil 2787; Nopcocide® N-40-D; Nopcocide® N-96; NuoCide®; Power Chlorothalonil 50; Siclor; Sipcam UK Rover 500; Tripart® Faber; Tripart® Ultrafaber; Pennington's Pride Multi-Purpose Fungicide; Pro-Care Multi-Purpose Fungicide; Rigo's Best Lawn & Garden Fungicide; SA Lawn Ornamental & Vegetable Fungicide; Security Fungi-Gard; Bravo-W-75; Dacobre; Echo 75; Vanox; Evade; BB Chlorothalonil. Chlorothalonil 500, 720 g/liter (40.4% w/w); a nonsystemic broad-spectrum fungicide for use in a wide variety of crops. Antimicrobial for marine antifouling coatings. mp = 250°; bp = 350°; d = 1.800; insoluble in H_2O (0.6 g/l), more soluble in organic solvents; LD_{50} (rat orl) = 10 gm/kg. *Farm Protection Ltd.; BASF plc; Chiltern Farm Chemicals Ltd.; Fermenta ASC Europe Ltd.; ICI Agrochemicals Professional Products; Henkel; Hüls Am.; Kommer-Brookwick Ltd.; Caffaro SpA; Sipcam UK Ltd.; Tripart Farm Chemicals Ltd.*

401 Copper Hydroxide
20427-59-2   2709   243-815-9
CuH_2O_2
Cupric hydroxide.
Comac Parasol; Chiltern Kocide 101; Kocide; Copper hydroxide; Comac parasol; Copper dihydroxide; Criscobre; Cudrox; Cuidrox; Cupravit blue;Kocide 101; Copper (II) hydroxide, Technical Grade, (stabilized), 96% (Assay). A protectant fungicide. *Chiltern Farm Chemicals Ltd.; McKechnie Chemicals Ltd.* See cupric hydroxide.

Fungicides

402 Copper Naphthenate
1338-02-9   215-657-0
Copper(II)naphthenate.
Naphthenic acids, copper salts; CNC; Copper uversol; Copper-nap-all; Cuprinol; Troysan; WILTZ-65; Wittox-C. Copper salt of petroleum naphthenic acids; wood, canvas and rope preservative; antifouling in paints; insecticide, fungicide. *Akzo; KMZ Chemical Ltd; Troy.*

403 Copper Oxychloride
1332-40-7   296-473-8
$CuCl_2.Cu_3H_6O_6$
Copper oxychloride.
dicopper chloride trihydroxide; copper(II) chloride oxide hydrate; Blitox; Ckuper; Cobox; Coprantol; Coptox; Criscobre; Cupravit; Cuprenox; Cuprocaffaro; Cuprokylt; Cuprosan; Cuprossina; Cuprovinol; Cuprox; Fytolan; Kauritil; Neoram; Recop; Viricuivre; Cupravit®; Vitigran; Cuprokylt; Cuprosana; Curenox-50; FS Dricol 50; Headland Inorganic Liquid Copper. Copper containing spray used for control of fungal diseases like downy mildews, late blight, early blight, apple scab and various leaf spot diseases on a wide range of crops. Insoluble in H_2O, organic solvents; LD_{50} (rat orl) = 1440 mg/kg. *Bayer AG; Cequisa; Ciba-Geigy; All-India Medical; Crystal; Diachem; Caffaro; Universal Crop Protection; Rhône Poluenc; Agrichem; Zeltia; BASF; Sandoz; Hoechst; Industrias Quimicas Del Valles SA; Ford Smith & Co Ltd; Bayer AG; Universal Crop Protection Ltd.; Industrias Quimicas Del Valles SA; WBC Technology Ltd.*

404 Croptex Fungex
33113-08-5   251-381-7
$C_2H_4CuN_2O_6$
Cupric ammonium carbonate.
Copper carbonate, basic; copper ammonium carbonate; carbonic acid, ammonium copper salt; Copper count N. Protectant fungicide. *Hortichem Ltd.*

405 CSC 2-Aminobutane
13952-84-6   1579   237-732-7
$C_4H_{11}N$
2-Aminobutane.
sec-butylamine; 2-Aminobutane; 2-Butylamine; 2-Butanamine; 2-aminobutane base; butafume; deccotane; frucote; Butanamine. Fungicide for stored potatoes. *Chemical Spraying Co Ltd.*

406 Cufraneb
11096-18-7
Ethylenebis(dithiocarbamate) complexed with 8.15% Zn, 8.05% Mn, 5.5% Cu, 1.0% Fe.
cufranebe. Foliar fungicide; acaricide. Used with together with cupric sulfate to control late blight in tomatoes and potatoes. LD_{50} (rat orl) = 2700 mg/kg.

407 Cupric Acetate, Basic
52503-64-7   2691   257-974-7
Cupric subacetate.
Complex of cupric acetate, cupric hydroxide, water; verdigris, green verdigris. Pig-ments, insecticides, fungicides, mold-preventatives.

408 Cupric Arsenate
7778-41-8
$As_2Cu_3O_8$
Copper(II)arsenate.
Cupar; Copper arsenate. Protective fungicide. *Mechema Chemicals Ltd.*

409 Cupric Arsenite
10290-12-7   2693   233-644-8
$CuHAsO_3$
Arsonic acid copper(2+) salt (1:1).
arsenious acid coppper(2+) salt(1:1); Scheele's green; pickle green. Used as a pigment, wood preservative, insecticide, fungicide, and rodenticide. Insoluble in H_2O, organic solvents.

410 Cupric Carbonate
12069-69-1   2697   235-113-6
$CuCO_3 \cdot Cu(OH)_2$
Basic Cupric carbonate.
malachite; [μ-[carbonato(2-)-O:O']]dihydroxydi copper; (carbonato(2-))dihydroxydicopper; copper carbonate hydroxide; cupric subcarbonate; Copper(II)hydroxide carbonate; Basic cupric carbonate; Bremen blue; Bremen green; Carbonato(2-(O:O'))dihydroxydicopper; Basic copper carbonate $(Cu_2(OH)_2CO_3)$. Used in pigments, pyrotechnics, insecticides, copper salts. An astringent in pomades, antidote for phosphorus poisoning, smut preventive, fungicide for seed treatment, feed additive. mp = 200° (dec); d = 4.000; LD_{50} (rat orl) – 1350 mg/kg. *Am. Chemet; Boliden Intertrade; Nihon Kagaku Sangyo.*

411 Cupric Sulfate, Basic
1332-14-5   2723   215-568-7
$Cu_4H_6O_{10}S$
Copper hydroxide sulfate.
Caldo Bordeles Valles; brochantite; langite; Cusatrib. Bordeaux mixture plus adjuvants; wettable powder used as protective fungicide for foliage application to ornamental and crop plants. Fungicide for plants, seed treatment. *Industrias Quimicas Del Valles SA; Mechema Chemicals Ltd.*

412 Cuprous Oxide
1317-39-1   2734   215-270-7
Cu_2O
Cuprous oxide.
Cupridan; Copper(I) oxide; Red Copper Oxide; C.I. 77402; Perenex; Perenox; Yellow Cuprocide; Copper-Sandoz; Caocobre; Cuprox; violet copper; Nordox; Copper oxide. Fungicide. Paint grade, red; active ingredient in antifouling paints. mp = 1232°; bp = 1800°; d = 6.000; LD_{50} (rat orl) = 0.47 g/kg. *Makhteshim Chemical Works Ltd.; Nordox Industrier AS; ICI Chem. & Polymers Ltd.*

413 Cymoxanil
57966-95-7   261-043-0
$C_7H_{10}N_4O_3$
2-Cyano-N-[(ethylamino)carbonyl]-2-(methoxyimino)acetamide.
1-(2-Cyano-2-methoxyiminoacetyl)-3-ethylurea; Curzate; DPX-3217. Foliar fungicide with protective and curative action. For control of Peronosporates, particularly *Peronospora, Phytophthora and Plasmopara* species. mp = 160-161°; d_{25} = 1.31; soluble in H_2O (1 g/l), freely soluble in organic solvents; LD_{50} (rat orl) = 1196 mg/kg. *DuPont.*

414 Dehydroacetic acid
520-45-6   2919   208-293-9
$C_8H_8O_4$
3-Acetyl-6-methyl-2H-pyran-2,4(3H)-dione ion(1-).
2H-pyran-2,4(3H)-dione, 3-acetyl-6-methyl-; 3-acetyl-4-hydroxy-6-methyl-2-pyrone cyclic ketone; DHA; DHAA; Methylacetopyronone. Fungicide, bactericide, plasticizer, chemical intermediate, medicated toothpastes. mp = 109-111°; bp = 269.9°; insoluble in H_2O, soluble in organic solvents; LD_{50} (rat orl) = 570 mg/kg.

415 Delan-Col
3347-22-6   3433   222-098-6
$C_{14}H_4N_2O_2S_2$
5,10-Dihydro-5,10-dioxonaphtho[2-3-b]-1,4-dithiin-2,3-dicarbonitrile.
Dithianone; Delan; Delan-col; DTA; MV 119A; Naphtho(2,3-b)-p-dithiin-2,3-dicarbonitrile, 5,10-dihydro-5,10-dioxo-; Stauffer MV-119a; Thynon. Suspension concentrate containing 600 g dithianon per liter; fungicide used for control of scab in fruit apples and pears. mp = 225°; insoluble in H_2O, soluble in organic solvents; LD_{50} (rat orl) = 638 mg/kg. *ICI Chem & Polymers Ltd.*

416 Dichlofluanid
1085-98-9   3095   214-118-7
$C_9H_{11}Cl_2FN_2O_2S_2$
1,1-Dichloro-N-[(dimethylamino)-sulfonyl]-1-fluoro-N-phenylmethanesulfenamide.

Euparen®; Bayer 47531; KUE 13032c; Elvaron®; Euparen (e); Euparen®. Fungicide with specific action against *Botrylis*. A wettable powder containing 50% w/w dichlofluanid; to control botrytis on berries, currants, outdoor grapes, tomatoes under cover, tulips and peonies; mp = 105-105.6°; insoluble in H_2O, soluble in organic solvents; LD_{50} (mus orl) = 1250 mg/kg. *Bayer AG.*

417 Dichlorophen
97-23-4   3120   202-567-1
$C_{13}H_{10}Cl_2O_2$
Bis(5-chloro-2-hydroxyphenyl)methane.
Algafen; Algofen; Dicestal; Ecco MP® 2004; Nuophene; Super Moss Killer & Lawn Fungicide; Super Mosstox; ; Prevental; Preventol; Preventol Gd; Preventol GDC; Taeniatol; Didroxan; Dichlorophen B; Wespuril; Super Mosstox; Gefir; G 4; Fungicide GM; Antiphen; Cuniphen; Dicestal; Dichlorophen; G-4 (Compound G4); Hyosan; Parabis; Teniathane; Teniatol; Westpuril; Diphenthane 70; Fungicide Fx; Korium; Panacide; Plath-lyse; Fungicide M; Dichlorofen; Antifen; Difentan; Teniotol; Gingivit; Cordocel; Halenol; Vermithana; Embephen; Palacel; DDDM; GH. A fungicide, bactericide, and algicide used as a moss-killer. A liquid formulation containing 34% dichlorophen; controls moss in fine turf, footpaths, hard tennis courts, playgrounds, roof and other affected hard surfaces. mp = 177-178°; insoluble in H_2O, soluble in organic solvents; LD_{50} (rat orl) = 1506 mg/kg. *Geeco; May & Baker Ltd.; Eastern Color & Chem.; Hüls Am.; Murphy Chemical Co Ltd.; Burts & Harvey; Rhône-Poulenc Environmental Prods. Ltd.*

418 Dicloran
99-30-9   202-746-4
$C_6H_4Cl_2N_2O_2$
2,6-Dichloro-4-nitrobenzeneamine.
Allisan; DCNA; Botran; Botran 75W; Dichloran; Dicloron; Ditranil; Kiwi Lustr 277; Resisan; RD-6584; AL-50; U-2069; CDNA; CNA; Resissan; Dichloro-4-nitroaniline; Fumite Dicloran. Horticultural fungicide containing dicloran. Protective fungicide used for control of *Botrytis, Monilinia, Rhizopus, Sclerotinia and Sclerotium* species in fruits and vegetables. mp = 195°; soluble in H_2O (6.3 mg/l), more soluble in organic solvents; LD_{50} (rat orl) = 4040 mg/kg. *The Boots Co plc; Octavius Hunt Ltd.*

419 Difolatan
2939-80-2   219-363-3
$C_{10}H_4Cl_4NO_2S$
cis-N-((1,1,2,2-Tetrachloroethyl)thio)-4-cyclohexene-1,2-dicarboximide.
Crisfolatan; Captafol; Difolatan; Folcid. Captafol fungicide. *Monsanto (Solaris).*

420 Fungicides

420 Dimethirimol
5221-53-4 3266 226-021-7
$C_{11}H_{19}N_3O$
5-Butyl-2-(dimethylamino)-6-methyl-4(1H)pyrimidinone.
dimethyrimol; Milcurb. Systemic fungicide with protective and curative action. Used as soil application for control of powdery mildews in curcubits, tobacco, capsicum, tomatoes and some ornamentals. mp= 102°; soluble in H_2O (1.2 g/l), more soluble in organic solvents; LD_{50} (rat orl) = 2350 mg/kg. ICI Chem & Polymers Ltd.

421 Diphenyl
92-52-4 3372 202-163-5
$C_6H_5C_6H_5$
1,1'-Biphenyl.
bibenzene; phenylbenzene; biphenyl. Used as a heat transfer agent, fungistat for agricultural use, in organic synthesis. Aldrich; Coalite Chem. Div; Koch; Monsanto; Sybron. See biphenyl.

422 Dodemorph-acetate
31717-87-0 250-778-2
$C_{20}H_{39}NO_3$
4-Cyclodecyl-2,6-dimethylmorpholine acetate.
4-cyclododecyl-2,6-dimethylmorpholine acetate; Meltatox; Meltatox®; Mehltaumittel; Milban; N-cyclododecyl-2,6-dimethylmorpholine acetate; F-238. Systemic fungicide with protective and curative action. Emulsifiable concentrate of 400g dodemorph per liter; used for control of mildew in ornamental nursery stock. mp = 63-64°; sparingly soluble in H_2O (<100 mg/l), more soluble in organic solvents; LD_{50} (rat orl) = 3944 mg/kg. BASF plc; BASF AG.

423 Dodine FL, WP
2439-10-3 3468 219-459-5
$C_{15}H_{33}N_3O_2$
Dodecylguanidine monoacetate.
Dodine; doguadine; AC 5223; Carpene; CL 7521; Curitan; Cyprex; Efuzin; Melprex; Radspor; Radspor FT, 65WP; Venturol; Dodine; AC-5223; Carpene; Cyprex; Melprex. A foliar fungicide with some protective and curative action; used for the control of scab in apples and pears. mp = 136°; soluble in H_2O (630 mg/l), EtOH, insoluble in organic solvents; LD_{50} (rat orl) = 100 mg/kg. Truchem Ltd.

424 Drazoxolan
5707-69-7 3499 227-197-8
$C_{10}H_8ClN_3O_2$
3-Methyl-4-[(2-chlorophenyl)hydrazone]-4,5-isoxazoledione.
PP-781; Ganocide; Mil-Col; Saisan. Mildew fungicide and seed dressing. mp = 168°; insoluble in H_2O, soluble in organic solvents; LD_{50} (rat orl) =

126 mg/kg. ICI Chem & Polymers Ltd; ICI Agrochemicals.

425 Ethirimol
23947-60-6 3786 245-949-3
$C_{11}H_{19}N_3O$
5-Butyl-2-(ethylamino)-6-methyl-4(1H)-pyrimidinone.
Milgo; Milstem; Milcurb Super. Fungicide containing ethirimol. Used as a seed dressing. ICI Chem & Polymers Ltd.

426 Etrimfos
38260-54-7 3936 253-855-9
$C_{10}H_{17}N_2O_4PS$
O-(6-Ethoxy-2-ethyl-4-pyrimidinyl) O,O-dimethyl phosphorothioate.
Satisfar; Ekamet; Ekamet G; Ekamet ULV; Etrimphos. Used to control pests in stored grain. Nickerson Seeds Ltd.

427 Euparen® M
731-27-1 211-986-9
$C_{10}H_{13}Cl_2FN_2O_2S_2$
1,1-Dichloro-N-((dimethylamino)sulfonyl)-1-fluoro-N-(4-methylphenyl)methanesulfonamide.
Tolylfluanid; BAY 49854; BAY 5712a; Dichlofluanid-methyl. Broad spectrum fungicide; effective against Botrytis. Bayer AG.

428 Fenaminosulf
140-56-7 205-419-4
$C_8H_{10}N_3NaO_3S$
p-Dimethylamino-benzenediazo sodium sulfonate.
Fenaminosulf, formulated; Formulated fenaminosulf; Methyl orange B; para-dimethylaminoazobenzenediazo sodium sulfonate; Bay 5072; Bay 72555; DAPA; Deksonal; Dexon; Dexoxon; Bayer 22555; diazoben; eniamethyl orange; gold orange hp; helianthin; Lesan; tropaeolin d; Bravo D; Bayer 5072. Fungicide for the prevention of crop damage by soil fungi. Bayer AG.

429 Fenpropidin
67306-00-7 4034
$C_{19}H_{31}N$
1-[3-[4-(1,1-Dimethylethyl)phenyl]-2-methyl propyl] piperidine.
Patrol; Ro-12-3049; Corbel®; Mistral; Power Task. Inhibitor of ergosterol biosynthesis. Fungicide. $bp_{0.2}$ = 117°; insoluble in H_2O. ICI Agrochemicals; BASF AG; BASF plc; Rhône-Poulenc Crop Protection Ltd.; Kommer-Brookwick Ltd.

430 Fluorfolpet
719-96-0 211-952-3
$C_9H_4Cl_2FNO_2$
N-(Fluordichloromethylthio) phthalimid.

N-(Dichlorofluoromethylthio)phthalimide; Di-chlorofluoromethyl)thio)phthalimide; Fluorofolpet; 2-((dichlorofluoromethyl)thio)isoindole-1,3(2H)-di-one; Preventol A3. Fungicide applied used in paints. *Bayer AG.*

431 Fluoromide
41205-21-4
$C_{10}H_4Cl_2FNO_2$
2,3-Dichloro-N-4-fluorophenylmaleimide.
Sparticide®. Foliar fungicide with protective action. Used for control of scab, *Alternaria* leaf spot and powdery mildew in apples, scab of citrus fruit and coffee berry disease. Fungicide for apple fruit spot, melanose, and scab of citrus, coffee berry disease, and pink disease on rubber. mp = 240-242°; soluble in H_2O (5.9 mg/l), slightly more soluble in organic solvents; LD_{50} (rat orl) > 15000 mg/kg. *Mitsubishi Kasei.*

432 Fluphenazine
76674-21-0   4226
$C_{22}H_{26}F_3N_3OS$
4-[3-[2-(Trifluoromethyl)-10H-phenothiazin-10-yl]propyl]-1-piperazineethanol.
Impact; Impact Excel; flutriafol; S-94; SQ-4918; 10-[3'-[4-(β-hydroxyethyl)-1-piperazinyl]propyl]-3-trifluoromethylphenothiazine. Fungicide. $bp_{0.5}$ = 268-274°. *ICI Chem & Polymers Ltd.*

433 Flusilazole
85509-19-9   4240
$C_{16}H_{15}F_2N_3Si$
1-[[Bis-(4-fluorophenyl)methylsilyl]methyl]-1H-1,2,4-triazole.
fluzilazol; DPX-H6573; Olymp; Punch; NuStar; Sanction. Foliar, systemic fungicide with protective and curative action. Used to control *Ascomycetes, Basdiomycetes* and *Deuteromycetes* in cereals, apples, vines and sugar beet. mp = 55°; soluble in H_2O (54 mg/l), more soluble in organic solvents; LD_{50} (rat orl) = 1110 mg/kg.

434 Folicur
107534-96-3   9253   403-640-2
$C_{16}H_{22}ClN_3O$
(±)-α-[2-(4-Chlorophenyl)ethyl]-α-(1,1-dimethylethyl)-1H-1,2,4-triazole-1-ethanol.
Tebuconazole; ethyltrianol; fenetrazole; terbuconazole; terbutrazole; BAY HWG 1608; HWG 1608; Corail; Elite; Folicur; Horizon; Lynx; Sivacur. Fungicide with systemic properties and broad spectrum activity against rusts, leaf spot diseases, e.g., *Septoria spp.*, powdery mildew and several *Fusarium* species on cereals; whitemold, Phoma and various leaf spot diseases on oilseed rape. *Bayer AG. See* tebuconazole.

435 Folpan
133-07-3   4255   205-088-6
$C_9H_4Cl_3NO_2S$
N-(Trichloromethylthio)phthalimide.
N-(trichloromethylmercapto)phthalimide; Phaltan; Phalton; Folpel; Folpex; Trichloromethyl(thio)-phthalimide. Agricultural fungicide. Used for control of downy mildews, powdery mildews, leaf spot diseases, scab, etc. in fruit and vegetable crops. mp = 177°; insoluble in H_2O, slightly soluble in organic solvents; LD_{50} (rat orl) > 10000 mg/kg. *Makhteshim Chemical Works Ltd.; Monsanto (Solaris).*

436 Fuberidazole
3878-19-1   223-404-0
$C_{11}H_8N_2O$
2-(2-Furanyl)-1H-benzimidazole.
Bay 33172; Voronit; W VII/117. Systemic fungicide used as a seed treatment for control of *Fusarium* spp. in cereals. mp = 284-288° (dec); soluble in H_2O (78 mg/l), more soluble in organic solvents; LD_{50} (rat orl) = 500 mg/kg. *Bayer AG.*

437 Furalaxyl
57646-30-7   260-875-1
$C_{17}H_{19}NO_4$
Methyl N-(2,6-dimethylphenyl)-N-(2-furanylcarbonyl)-DL-alanine.
Fongarid; furalaxyl; Fonganil; CGA-38140. Systemic fungicide with protective and curative action for ornamentals. Used for control of soil diseases caused by *Phytophthora and Pythium* spp. mp = 70°, 84°; d^{20} = 1.22; soluble in H_2O (230 mg/l), more soluble in organic solvents; LD_{50} (rat orl) = 940 mg/kg. *Ciba-Geigy Agrochemicals.*

438 Guanoctine
108173-90-6
$C_{10}H_{24}N_6$
Guazatine.
GTA; Rappor; EM 379; Kenopel; MC 25; Panoctine; Panolil; Kenogard. A fungicide seed dressing for wheat. Has bird-repellent properties. *DowElanco Ltd.; KenoGard; Rhône-Poulenc.*

439 Hinosan®
17109-49-8   3560   241-178-1
$C_{14}H_{15}O_2PS_2$
O-Ethyl S,S-diphenyl phosphorodithioate.
Edifenphos; EDDP; Bay 78418; Blastoff; Hinosan. Foliar fungicide especially effective against *Pyricularia oryzae* on rice; also for control of other rice diseases. $bp_{0.01}$ = 154°; d^{20} = 1.23; n_D^{22} = 1.6112; soluble in H_2O (56 mg/l), more soluble in organic solvents; LD_{50} (rat orl) = 212-240 mg/kg. *Bayer AG. See* edifenphos.

440　　Imazalil

35554-44-0　　　　　3522　　　　252-615-0

$C_{14}H_{14}Cl_2N_2O$

(±)-1-[2-(2,4-Dichlorophenyl)-2-(2-propenyloxy)-ethyl]-1H-imidazole.

Fungaflor; enilconazole; chloramizol; Bromazil; Deccozil; Fecundal; Florasan; Freshguard; Fungaflor; Fungazil; R 23979; Clinafarm; Magnate; Rappor Plus; Magnate. Ergosterol biosynthesis inhibitor, used as a systemic fungicide with protective and curative action. Used to control fungal diseases in fruit, vegetables and ornamentals. Used for control of mildew in greenhouse plants. mp = 50°; SG_{23} = 1.243; n_D^{20} = 1.5643; soluble in H_2O (1.4 g/l), more soluble in organic solvents; LD_{50} (rat orl) = 320 mg/kg. *Hortichem Ltd; Janssen Pharmaceutical Ltd; Makhteshim Chemical Works Ltd; Dow; Elanco Ltd.*

441　　Iprodione

36734-19-7　　　　　5093　　　　253-178-9

$C_{13}H_{13}Cl_2N_3O_3$

3-(3,5-Dichlorophenyl)-N-(1-methylethyl)-2,4-dioxo-1-imidazolidinecarboxamide.

CDA Roval; Roval Dust; Roval Flo; Roval Green; Roval WP; Turbair Roval. A fungicide with protectant activity for use in turf, amenity grasses, field crops, cereals, fruit trees, bulbs, lettuce, nursery stock, fruit trees and glass house crops. *Rhône-Poulenc Environmental Prods. Ltd.; Rhône-Poulenc Crop Protection Ltd.; Hortichem Ltd.; Embetec Crop Protection Ltd.; Pan Britannica Industries Ltd.*

442　　Kocide® Copper Sulfate Pentahydrate Crystals

7758-99-8　　　　　2722　　　　231-847-6

$CuSO_4.H_2O$

Cupric sulfate pentahydrate.

Copper(II) sulfate pentahydrate; cupric sulphate, pentahydrate; bluestone; blue vitriol; Roman vitriol; Salzburg vitriol. Fungicide to control plant diseases, in fertilizers to correct copper deficiencies in soils. Becomes anhydrous by 250°; $d^{15.6}_4$ = 2.286. *Griffin.*

443　　Korlan

299-84-3　　　　　8415　　　　206-082-6

$C_8H_8Cl_3O_3PS$

Ronnel.

Phosphorothioic acid, O,O-dimethyl O-(2,4,5-trichlorophenyl)ester; fenchlorphos; dimethyl trichlorophenyl thiophosphate; Trolene; Etrolene; Nankor; Viozene; Ectorl. Active ingredient: ronnel; insecticide used on cattle for the control of ticks, files, maggots, and lice. mp = 41°. *Dow UK.*

444　　Krystallazurin

14283-05-7　　　　　9323　　　　238-177-3

$CuH_{12}N_4O_4S$

Tetraamminecopper sulfate.

Cuprammonium sulfate; ammonium cupric sulfate; cupric sulfate, ammoniated. Fungicide consisting of ammoniacal copper sulfate. d_4^{20} = 1.81.

445　　Kumulus® S

7704-34-9　　　　　9142　　　　231-722-6

S

Sulfur.

Sulphur; brimstone; Vassgro Flowable Sulphur; Gofrativ; Gofravik; Green Sulphur; Solfa; Hortag Aquasulf; Stoller Flowable Sulphur; Thiovit; Crude Sicilian sulfur; soufre; Alfa, Aquilite; Cosan; Elosal; Golden Dew; Imber; Kolodust; Kolofog; Kolospray; Kumulus; Magnetic 6; Solfa; Suffa; Sulfex; Sulflox; Sulfospor; Sulphotox; Super Six; That; Thiolux; Thion; This; Tiolene; Uniflow; Zolvis. Wettable powder containing 80% w/w sulfur per liter; a protectant fungicide/foliar feed. Used for mildew control in a wide range of crops and for control of scab in fruits and mildews on a variety of crops. Also used as an acaricide. mp = 117-120°; bp = 444°; d = 2.060. *BASF plc; BASF AG; L W Vass (Agricultural) Ltd.; Makhteshim Chemical Works Ltd.; Vitax Ltd.; Avon Packers Ltd.; ICI Chem & Polymers Ltd.; Farm Protection Ltd.; Stoller Chemicals Ltd.; Draxel Chemical Company; Pan Britannica Industries Ltd.*

446　　Mackechnie

7758-98-7　　　　　2722　　　　231-847-6

Crystallized copper sulfate; Sulfato de Cobre Valles. Copper sulfate, pentahydrate, monohydrate and anhydrous. Used as a fungicide and in manufacture of agricultural fungicides and many industrial products. *McKechnie Chemicals Ltd,; Industrias Quimicas Del Valles SA. See cupric sulfate.*

447　　Mancozeb

8018-01-7　　　　　5756

$[C_4H_6N_2S_4Mn]_x[Zn^{2+}]_y$ where x:y = 10:1

Manganese (II) ethylenebis (dithiocarbamate).

[[1,2-Ethanediylbis(carbamodithioato)](2-)]manganese, mixture with [[1,2-ethanediylbis-(carbamodithioato)](2-)]zinc; ethylenebis(dithiocarbamic acid) manganese zinc complex; manzeb; manganese ethylenbis(dithiocarbamate) (polymeric) complex with zinc salt; zinc manganese ethylenebisdithiocarbamate; Dithane; Dithane M-45; Manzate; Manzate® 200 DF; Manzin; Nemispor; Penncozeb; Vondozeb; Manplex; Penncozeb; Karamate. Wettable powder or water dispersible granules containing mancozeb; protectant fungicide for fruit, field crops, and roses. Used for control of potato blight and rust, blight and mildew in winter wheat.

Rohm & Hass UK; Pan Britannica Industries Ltd.; Agrimont; Akzo; All-India Medical; Crystal; Desarrollo Quimica; Diachem; DuPont; Ercros; Pennwalt Holland; Rohm & Haas, Sanachem; Kommer-Brookwick Ltd.; DuPont UK; Shell UK.

448   Maneb
12427-38-2   5761   235-654-8
$C_4H_6MnN_2S_4$
[[1,2-Ethanediylbis[carbamodithioato]](2-)] manganese.
[ethylenebis[dithiocarbamato]]manganese; m-Diphar; Amangan; ethylenebis[dithiocarbamic acid manganese salt; Chem neb; Chloroble B; CR 3029; Dithane M 22; Dithane M 22 special; Dithane M-45; Dithane S-31; Campbell's X-Spor; Headland Spirit; Manguard®; Manzate®; Septal; Trimangol; EBDC, manganese salt; F 10; Griffin Manex; Kypman 80; Labilite; Lonocol M; M-Diphar; Manam; 249; Maneb 80; Maneb-R; Maneba; manebe; MEB; ENT 14875;Kypman; Man-Zox; Manex; Manox; Manzi; Polyram M; Trimangol. A dithiocarbamate fungicide with protective action. Used in control of many fungal diseases, e.g. blight, leaf spot, rust, downy mildew, scab, etc. Suspension concentrate containing 480 g maneb per. mp = 192-204°; insoluble in H_2O and common solvents; LD_{50} (rat orl) = 3000 mg/kg. Rohm & Haas; Cumberland; Crystal; Drexel; BASF; Chiltern Farm Chemicals Ltd; Dupont; Pennwalt Holland; MTM Agro-Chemicals Ltd.; WBC Technology Ltd.; Universal Crop Protection Ltd.; DuPont UK; Schering Agrochemicals Ltd.; Pennwalt Chemicals Ltd.

449   Maxicrop Moss Killer & Conditioner
10028-22-5   4079   233-072-9
$Fe_2O_{12}S_3$
Iron (III) sulfate.
Ferric sulfate; ferric persulfate; ferric sesquisulfate; ferric tersulfate; Sulfuric Acid, Iron(3+) Salt (3:2); Iron Tersulfate; Ferric sulfate monohydrate. Used for moss control in turf. d^{18} = 3.097; slowly soluble in H_2O, EtOH, poorly soluble in organic solvents. Maxicrop International Ltd.; Vitax Ltd. See ferric sulfate.

450   MCPB
94-81-5   202-365-3
$C_{11}H_{13}ClO_3$
3-Phenoxybutyric acid.
Tropotox; Can-Trol; PDQ; 2,4-MCPB; Bexane; Mcp-butyric; Trifolex; MCPB - acid. Used as a systemic fungicide active against chocolate spot disease in broad beans.

451   Metalaxyl
57837-19-1   5982   260-979-7
$C_{15}H_{21}NO_4$
N-(2,6-Dimethylphenyl)-N-(methoxyacetyl)-DL-alanine methyl ester.
metaxanin; CGA-48988; Apron; Ridomil; Subdue. Fungicide. mp = 71-72°; soluble in H_2O (7.1 g/l), organic solvents; LD_{50} (rat orl) = 669 mg/kg.

452   Metiram
9006-42-2
Tris(amine)(ethylenebis(dithiocarbamato))zinc(2+)(tetrahydro-1,2,4,7-dithiadiazocene-3,8-dithione) polymer.
Polyram®; ammoniated EBDCs; Polyran; Carbamodithioic acid; Carbatene; Ethylenebis-(dithiocarbamic acid) polymer with ammonia complex of zinc ebdc; Polyram-combi; Zinc metiram. Prevents crop damage in the field, during storage, or transport. Effective against a broad spectrum of fungi and is used to protect fruits, vegetables, field crops, and ornamentals from foliar diseases and damping off. Practically non-toxic. mp = 140°; insoluble in H_2O. BASF AG.

453   Mirvale
101-21-3   2240   202-925-7
$C_{10}H_{12}ClNO_2$
Isopropyl m-chlorocarbanilate.
Warefog; Mirvale; Furloe; Chloro-IPC; Isopropyl N-(3-chlorophenyl) carbamate; Spud-Nic; Taterpex; CIPC; Sprout Nip; Chloropropham; (3-chlorophenyl)carbamic acid 1-methylethyl ester; MSS CIPC; N-3-Chlorophenylisopropylcarbamate; Furloe Chloro IPC 4EC; Furlow Chloro IPC 20G; Beet; Bud-Nip; Beet-Kleen; Chlorocarbanilic acid isopropyl ester; Elbanil; Fasco WY-HOE; Jack Wilson chloro 51 oil; Liro cipc; Nexoval; Preweed; Spud-nip; Stopgerme-S. Potato sprout depressant. Supplied as a foggable solution; for controlling sprouting in ware potatoes. A carbamate herbicide and sprout depressant in stored potatoes. Ciba-Geigy Agrochemicals; Wheatley Chemical Co Ltd; Dean Agrochemicals Ltd; Mirfield Sales Services Ltd. See chloropropham.

454   Morocide
485-31-4   1265   207-612-9
$C_{14}H_{16}N_2O_6$
2-(1-Methylpropyl)-4,6-dinitrophenyl 3-methyl-2-butenoate.
binapacryl; Acricid; Ambox; Dapacryl; Endosan; FMC 9044; Hoe 002784; Morrocid; NIA 9044. Fungicide for mildew. Hoechst UK.

455   Mozanon®
9005-38-3   240
Sodium alginate
Algin. Fungicide for prevention of TMV infection on tobacco. Mitsubishi Kasel. See sodium alginate.

456   Myacide® SP
1777-82-8   3110   217-210-5
$C_7H_6Cl_2O$
2,4-Dichlorobenzyl alcohol.
Dybenal. Antifungal agent, preservative. Inolex; *Boots Co. PLC. See* 2,4-dichlorobenzyl alcohol.

457   Myclobutanil
88671-89-0   6402
$C_{15}H_{17}ClN_4$
α-Butyl-α-(4-chlorophenyl)-1,2,4-triazole-1-propanenitrile.
RH-3866; Nova; Rally; Systhane. Systemic fungicide with protective and curative action. Used in control of Ascomycetes, Fungi Imperfecti and Basidomycetes in a wide variety of crops. mp = 63-68°; bp_1 = 202-208°; soluble in H_2O (142 mg/ml), polar organic solvents; LD_{50} (rat orl) = 1600 mg/kg. *Rohm & Haas; Hoechst UK; Pan Britannica Industries Ltd; Rohm & Haas UK.*

458   Nitrothal-Isopropyl
10552-74-6   234-139-5
$C_{15}H_{19}NO_6$
Bis(1-methylethyl) 5-nitro-1,3-benzenedicarboxylate.
nitrothale-isopropyl; BAS 30000F. Non-systemic contact fungicide with curative action. Used in combination with other fungicides to control powdery mildews on apples, vines, hops, vegetables and ornamentals. mp = 65°; poorly soluble in H_2O (0.39 mg/l), more soluble in organic solvents; LD_{50} (rat orl) > 6400 mg/kg.

459   Nuarimol
63284-71-9   264-071-1
$C_{17}H_{12}ClFN_2O$
α-(2-Chlorophenyl)-α-(4-fluorophenyl)-5-pyrimidinemethanol.
Cidorel; EL-228; Gandural; Gauntlet; Murox; Nuarimol; Tridal; Trimidal; Triminol. Systemic foliar fungicide with curative and protective action. Ergosterol biosynthesis inhibitor. Used for control of a wide range of pathogenic fungi. Used as a foliar spray or seed treatment. Emulsifiable concentrate containing 90 g/l nuarimol; pyrimidine fungicide to control mildew. mp = 126-127°; soluble in H_2O (26 mg/l), more soluble in organic solvents; LD_{50} (rat orl) = 1250 mg/kg. *DowElanco Ltd.*

460   Octhilinone
26530-20-1   6853   247-761-7
$C_{11}H_{19}NOS$
2-Octyl-4-isothiazolin-3-one.
Pancil T; Kathon 893; RH 893; Skane M-8; kathon lp preservative; kathon sp 70; micro-chek 11; micro-chek 11d; micro-chek skane; pancil; pancilt; skane hq; Kathon; Microbicide M-8. Paste containing 1% w/w octhilinone; used as a fruit tree canker paint. *Rohm & Haas UK. See* octhilinone.

461   Orthorix
1344-81-6   215-709-2
CaS_x
Calcium polysulfide.
lime-sulfur; Eau grison; Neviken; Security Lime Sulphur. Lime-sulfur fungicide. Acts directly and alo by decomposition to sulfur which is also a fungicide. Used for control of powdery mildews, anthracnose, scab and other diseases in benas, clover and fruits. Control of insects and spider mite eggs in fruit trees. SG15.6:kls > 1.28; soluble in H_2O. *Monsanto (Solaris).*

462   Oruface
58810-48-3   261-451-9
$C_{14}H_{16}ClNO_3$
2-Chloro-N-(2,6-dimethylphenyl)-N-(tetrahydro-2-oxo-3-furanyl)acetamide.
milfuram; Patafol. Fungicide containing ofurace for the control of potato blight. *(Sold in UK for Chevron Chemical Co) . ICI Chem & Polymers Ltd.*

463   Oxadixyl
77732-09-3
$C_{14}H_{18}N_2O_4$
N-(2,6-Dimethylphenyl)-2-methoxy-N-(2-oxo-3-oxazolidinyl)acetamide.
Anchor; SAN 371F; Sandofan. Systemic fungicide with protective and curative action. Used in combination with contact fungicides (e.g. mancozeb; captofol etc.) for control of downy mildews, late blights and rusts in vines, potatoes, maize, tobacco, hops, sunflowers, citrus, fruits and vegetables. mp = 104-105°; soluble in H_2O (3.4 g/kg), more soluble in organic solvents; LD_{50} (rat orl) = 3480 mg/kg.

464   Oxycarboxin
5259-88-1   226-066-2
$C_{12}H_{13}NO_4S$
5,6-Dihydro-2-methyl-N-phenyl-1,4-Oxathiin-3-carboxamide 4,4-dioxide.
Plantvax; Carboxin sulfone; DCMOO; Oxykisvax; Ringmaster. A systemic fungicide with curative action. Used for the control of rust diseases in ornamentals, nursery trees and wheat. mp = 127-130°; soluble in H_2O (1 g/l), more soluble in organic solvents; LD_{50} (rat orl) = 2000 mg/kg. *ICI Agrochemicals; Uniroyal Chemical Ltd.; Fargro Ltd.; Rhône-Poulenc Environmental Prods. Ltd.*

465   P.N.P.
100-02-7   6718   202-811-7
$C_6H_5NO_3$
p-Nitrophenol.

4-Hydroxynitrobenzene; Niphen; PNP; mononitrophenol; 4-Nitrophenol. Used as fungicide in the rubber industry.

466   Panogen M
151-38-2   205-790-2
$C_5H_{10}HgO_3$
2-Methoxyethylmercury acetate.
(acetato-O)(2-methoxyethyl)mercury; methoxyethyl mercury acetate; MEMA; Panogen; acetoxy(2-methoxyethyl)mercury. Cereal seed treatment. Formulated as aqueous solutions or dusts and used chiefly as a seed protectant. Use of alkyl mercury fungicides in the United States has been virtually prohibited for several years. Phenyl mercuric acetate is still used to control diseases of turf, but other applications have been sharply restricted. *Embetec Crop Protection Ltd.*

467   Pencycuron
66063-05-6   266-096-3
$C_{19}H_{21}ClN_2O$
N-((4-Chlorophenyl)methyl)-N-cyclopentyl-N'-phenylurea.
Bay NTN 19701; Monceren; [(chlorophenyl)-methyl]-N-cyclopentyl-N'-phenylurea; Trotis. Dustable powder containing 12.5% w/w pencycuron; a phenylurea fungicide used to control black scurf in potatoes. Non-systemic fungicide used fro control of *Rhizoctonia and Pellicularia* spp. in potatoes, rice, cotton and vegetables. In particular, control of black scurf of potatoes, sheath blight of rice, and damping-off of ornamentals. mp = 129°; insoluble in H_2O (0.4 mg/l), soluble in organic solvents; LD_{50} (rat orl) > 5000 mg/kg. *Bayer plc.*

468   Plondrel
5131-24-8   225-875-8
$C_{12}H_{14}NO_4PS$
O,O-Diethyl (1,3-dihydro-1,3-dioxo-2H-isoindol-2-yl)phosphonothioate.
Diethyl phthalimidophosphonothioate; Dowco 199; Laptran; Plondrel. Fungicide containing ditalimfos; for the control of powdery mildew and scab. *(Sold in UK on behalf of Dow Chemical Co). ICI Chem. & Polymers Ltd.*

469   Potassium Benzoate
582-25-2   1122   209-481-3
$C_7H_5KO_2.3H_2O$
Benzoic acid potassium salt trihydrate.
Anti-corrosive, preservative, fermentation-inhibitor, anti-fungal agent used in tobacco production, pyrotechnical additive. Soluble in H=72O, EtOH. *Am. Biorganics; Mallinckrodt; Pentagon Chemicals Ltd; Schweizerhall; Verdugt BV.*

470   Potassium Sorbate
24634-61-5   7841   246-376-1
$CH_3CH.cCHCH.cCHCOOK$
2,4-Hexadienoic acid potassium salt.
sorbic acid, potassium salt; potassium 2,4-hexadienoate. Used as a mold and yeast inhibitor. mp = 270° (dec); d_{20}^{25} = 1.363; soluble in H_2O (58 g/100 ml), less soluble in organic solvents. *Chisso Am.; Gist-Brocades Food Ingreds.; Hoechst Celanese; Pfizer Spec. Chem.; Protameen.*

471   Prochloraz
67747-09-5   7941   266-994-5
$C_{15}H_{16}Cl_3N_3O_2$
N-Propyl-N-[2-(2,4,6-trichlorophenoxy)ethyl]-1H-imidazole-1-carboxamide.
Mirage; Octave; Prelude; Sporak; Sporgon; Sportak Delta; Ascurit; BTS 40542; Octave; Omega; Prelude; Sporgon; Sportak; cyproconazole. Wettable powder containing 50% w/w prochloraz; a broad-spectrum fungicide for cereal crops, used to protect against seed-borne diseases. Applied as a foliar spray to control *Rhynchosporium, Helminthosporium, Septoria, Fusarium, Pseudocercosporella, Erysiphe and Pyrenophora* species in crops. *Makhteshim Chemical Works Ltd.; Fisons plc; Agrichem (International) Ltd.; Darmycel UK; Schering Agrochemicals Ltd.*

472   Propamocarb Hydrochloride
25606-41-1   247-125-9
$C_9H_{21}ClN_2O_2$
Propyl (3-(dimethylamino)propyl)carbamate monohydrochloride.
Filex; Banol Turf Fungicide; Prevex; Previcur N; Propamocarb hydrochloride. Aqueous concentrate containing propamocarb hydrochloride; a protective fungicide for use on all ornamentals and some edible crops against *Pythium, Peronospora and Phytophthora. Fisons plc.*

473   Propiconazole
60207-90-1   8003   262-104-4
$C_{15}H_{17}Cl_2N_3O_2$
1-[[2-(2,4-Dichlorophenyl)-4-propyl-1,3-dioxolan-2-yl]methyl]-1H-1,2,4-triazole.
Propiconazole; CGA-64250; proconazole; Banner; Bumper; Desmel; Orbit; Tilt; Power Propiconazole; Powerspire; Radar; Radar Propiconazole. A systemic triazole fungicide for control of powdery mildew and rust in cereals. $bp_{0.1}$ = 180°; soluble in H_2O (110 mg/l), more soluble in most organic solvents; LD_{50} (rat orl) = 1517 mg/kg. *Makhteshim Chemical Works Ltd.; Kommer-Brookwick Ltd.; Farm Protection Ltd; Ciba-Geigy Agrochemicals; ICI Chem. & Polymers Ltd.*

Fungicides

474

474 Propineb

12071-83-9   8004   235-134-0

$C_5H_8N_2S_4Zn$

Zinc (N,N'-propylene-1,2-bis(dithiocarbamate)).
Antracol®; Fruvit®; Airone; zinc propylenebis-(dithiocarbamate); mezineb; Antracol. Fungicide for protective control of potato blight, hop downy mildew, apple scab, leafspot on celery, blackcurrants and gooseberries, downy mildew on grapes and suppression of yellow rust on winter wheat. *Bayer AG.*

475 Propylparaben

94-13-3   8051   202-307-7

$C_{10}H_{12}O_3$

Propyl p-hydroxybenzoate.
4-hydroxybenzoic acid propyl ester; propyl parahydroxybenzoate. Organic ester of n-propyl alcohol and p-hydroxybenzoic acid; food preservatives, fungicide, mold control in sausage casings, pharmaceutic aid. *Allchem Industries; R.W. Greef; Mipa Labs; Penta Mfg.*

476 Quintozene

82-68-8   8264   201-435-0

$C_6Cl_5NO_2$

Pentachloronitrobenzene.
Botrilex; Brabant PCNB; Folosan; Tubergran; Avicol; Botrilex; PCNB; PKhNB; Brassicol; Earthcide; Folosan; Kobu; Kobutol; Pentagen; Saniclor; Terraclor; Terrazan; Tilcarex; Tritisan; Tubergran; Turfcide. A horticultural fungicide containing quintozene. A contact fungicide containing 50% w/w quintozene; used for the control of diseases in turf. Seed and soil-fungicide used for control of fungal diseases in fruits and vegetables. mp = 143-144°; bp = 328° (dec); insoluble in H_2O (0.44 mg/l), soluble in organic solvents; LD_{50} (rat orl) > 12000 mg/kg. *ICI Chem & Polymers Ltd.; Bos Chemicals Ltd.; Burts & Harvey; Rhône-Poulenc Environmental Prods. Ltd.; Wheatley Chemical Co Ltd.*

477 Sodium Benzoate

532-32-1   8725   208-534-8

$C_7H_5NaO_2$

Benzoic acid sodium salt.
Sodium salt of benzoic acid; fungicide; preservative in pharmaceuticals and foods, especially in slightly acidic media; clinical reagent (bilirubin assay). Soluble in H_2O (555 mg/ml); LD_{50} (rat orl) = 4.07 g/kg. Aceto; Dinoval Chem. Ltd; *DSM BV; Haarmann & Reimer; Pentagon Chemicals Ltd; Mallinckrodt.*

478 Sodium Dimethyldithiocarbamate

128-04-1   204-876-7

$C_3H_6NNaS_2$

Sodium N,N-dimethyl dithiocarbamate.
SDDC; dimethyldithiocarbamic acid sodium salt; methyl namate; sodium dimethyldithiocarbamate; Aceto SDD 40; Alcobam NM; Brogdex 555; Carbon S; Dibam; Dibam A; DMDK; Sharstop 204; sodium N,N-dimethyldithiocarbamate; Stafresh 615; Steriseal 40; Thiostop N; Vinstop; Vulnopol NM; Wing stop b; SDMDTC; sodium dimethylcarbamodithioate; Dimethyldithiocar-bamic acid sodium salt; Freshgard 40; Novate SM-40. Pesticide, fungicide, corrosion inhibitor, rubber accelerator. [trihydrate]: mp = 94-102°; soluble in H_2O, polar organic solvents; λ_m = 257, 290 nm (ε 1200, 13000 EtOH); LD_{50} (rat orl) = 2830 mg/kg. *NovaChem; Uniroyal; R. T. Vanderbilt Co Inc.*

479 Sodium Fluoride

7681-49-4   8762   231-667-8

FNa

Disodium difluoride; Floridine; Florocid; Villiaumite; NaF; sodium hydrofluoride; sodium monofluoride; trisodium trifluoride; alcoa sodium fluoride; antibulit; cavi-trol; chemifluor; Credo; duraphat; fda 0101; f1-tabs; flozenges; fluoral; fluorident; fluorigard; fluorineed; fluorinse; fluoritab; fluorocid; fluor-o-kote; fluorol; fluoros; Flura; flura-gel; flura-loz; flurcare; flursol; fungol b; Gel II; gelution; Gleem; iradicav; karidium; karigel; kari-rinse; lea-cov; lemoflur; luride; luride lozi-tabs; luride-sf; nafeen; nafpak; na frinse; nufluor; ossalin; Ossin; osteofluor; pediaflor; pedident;pennwhite; pergantene; phos-flur; point two; predent; raflour; rescue squad; Roach salt; sodium fluoride cyclic dimer; So-flo; stay-flo; studafluor; super-dent; t-fluoride; thera-flur; thera-flur-n; zymafluor; Les-cav. Fluoridation of municipal water, degassing steel, wood preservative, insecticide, fungicide, rodenticide, chemical cleaning, electroplating, glass manufacture, vitreous enamels, preservative for adhesives, toothpastes, disinf toothpastes, disinfectants, dental prophylaxis. mp = 993°; bp = 1704°; d = 2.78; soluble in H_2O (4 g/100 ml); insoluble in organic solvents; LD_{50} (rat orl) = 0.18 g/kg. *Cerac; EM Industries; General Chem; Hoechst Celanese; Solvay GmbH; Whiting, Peter Ltd.*

480 Sorbistat

110-44-1   8869   203-768-7

$C_6H_8O_2$

Sorbic acid.
1,3-pentadiene-1-carboxylic acid; Preservastat; Sorbistat; hexadienoic acid; trans,trans-sorbic acid; alpha-trans-gamma-trans-sorbic acid; trans,trans-2,4-hexadienoic acid; (E,E)-2,4-hexadienoic acid; panosorb; (2-butenylidene)acetic acid; crotylidene acetic acid; Hexadienoic acid, (E,E); hexa-2,4-dienoic acid. Fungistatic agent, used on cheeses. *Pfizer Ltd. See sorbic acid.*

481 Sorbistat K

590-00-1

$C_6H_7KO_2$

Potassium sorbate.

2,4-Hexadienoic acid, potassium salt; Hexadienoic acid, potassium salt. Fungicide. *Pfizer Ltd. See* potassium sorbate.

482 Sulfasan

103-34-4   3437   203-103-0

$C_8H_{16}N_2O_2S_2$

4,4'-Dithiodimorpholine.

4,4'-dithiobis[morpholine]; morpholine, N,N'-disulfide; dimorpholine N,N'-disulfide; Sulfasan R. Vulcanizing agent for natural and synthetic rubbers. Also used as a fungicide. mp = 124-125°. *Monsanto Co.*

483 Tachigaren 70

10004-44-1   4905   233-000-6

$C_4H_5NO_2$

5-Methyl-3(2H)-isoxazolone.

Hymexazol; 3-hydroxy-5-methylisoxazole; F-319; RTY-319. Fungicide for pelleting sugar beet seed. mp = 84-85°; soluble in H_2O (8.5 g/100 ml), very soluble in organic solvents; LD_{50} (rat orl) = 4678 mg/kg. *Sumito Chemical (UK) plc.*

484 Tebuconazole

107534-96-3   9253   403-640-2

$C_{16}H_{22}ClN_3O$

(RS)-1-(4-Chlorophenyl)-4,4-dimethyl-3-(1H-1,2,4-triazol-1-ylmethyl)pentan-3-ol.

Raxil; Silvacur; Horizon/Horizont. Fungicide with systemic properties/broad spectrum activity against rusts, leaf spot diseases, e.g., *Septoria spp.*, powdery mildew and several Fusarium species on cereals; whitemold, *Phoma* and various leaf spot diseases on oilseed. Cereal seed dressing with systemic properties for control of seed-borne diseases such as stinking smut, loose smuts and covered smut; highly effective at low dosage rates. mp = 104-107°; soluble in H_2O (32 mg/l), more soluble in organic solvents. *Bayer AG.*

485 Thiabendazole

148-79-8   9426   205-725-8

$C_{10}H_7N_3S$

2-(4-Thiazolyl)-1H-benzimidazole.

Thiabendazole; benzimidazole, 2-(4-thiazolyl)-; Apl-Luster; Arbotect; Bioguard; Bovizole; Eprofil; Equizole; Lombristop; Mertec; Mertect; Mertect 160; Metasol TK 100; Mintesol; Mintezol; Minzolum; Mycozol; MK 360; Nemapan; Omnizole; Polival; Tbz; Tebuzate; Tecto; Tecto RPH; Tecto 10P; Tecto 40F; Tecto 60; Testo; Thiaben; Thiabendazol; Thiprazole; Tiabenda; Hymush; Nemacin; Tecto; Tecto 60%; Tubazole. Systemic fungicide with protective and curative action. Absorbed by leaves and roots; used for control of fungus in vegetabls, fruits and cereals. Thiabendazole and iodophors such as Dermevan, Idonyx or Wescodyne as a foggable solution; used for controlling various fungal diseases in stored potatoes. mp = 300° (dec); soluble in H_2O (250 mg/l at pH 2-5), more soluble in organic solvents; LD_{50} (rat orl) = 2080 mg/kg. *MSD Agvet; Pennwalt; Duphar; Agrichem; Agrichem (International) Ltd.; BASF, Ciba-Geigy; Dow Elanco; Wheatley Chemical Co Ltd.*

486 Thiophanate-methyl

23564-05-8   9489   245-740-7

$C_{12}H_{14}N_4O_4S_2$

Dimethyl ester 4,4'-o-Phenylenebis-(3-Thioallophanic) Acid.

Sys Tec® 1998; Thiophanate-Me; Dragon Systemic Fungicide 3336WP; Ferti-lome Halt Systemic; Green Light Systemic Fungicide; SA Thiomyl Systemic Fungicide; Bas 32500f; caligran; cercobin m; cercobin m 70; cercobin methyl; cycosin; asout; enovit m; enovit m 70; enovit methyl; enovit super; enovit-supper; F 6385; frumidor; Fungo; fungo 50; labilite; metoben; methylthiofanate; methylthiophanate; methyl topsin; mildothane; neotopsin; NF 44; pei 190; pelt 14; pelt-44; Sigma; sipcaplant; sipcasan; sipcavit; td 1771; topsin m; topsin m 70; topsin nf-44; topsin wp methyl; trevin; Zyban; Spot Kleen. A broad-spectrum systemic fungicide and wound protectant which controls a variety of diseases for use on turf and ornamentals. *Regal Chemical Company.*

487 Thiotax

149-30-4   5916   205-736-8

$C_7H_5NS_2$

2(3H)-Benzothiazolethione.

2-mercaptobenzothiazole; 2-benzothiazolethiol; MBT; Captax; Dermacid; Mertax. Vulcanization accelerator. Zinc and sodium salts used as a fungicide. mp = 179°; d = 1.4200; insoluble in H_2O, soluble in organic solvents. *Monsanto Co.*

488 Thiram

137-26-8   9510   205-286-2

$C_6H_{12}N_2S74$

Tetramethylthiuram disulfide.

bis(dimethylthiocarbamyl) disulfide; TMTD; tetramethylthioperoxydicarbonic diamide; ENT 987; SQ 1489; NSC 1771; Unicrop Thianosan; Arasan; Thiurad; Thiosan; Thylate; Tiuramil; Thiuramyl; Puralyn; Fernasan; Nomersan; Rezifilm; Pomarsol; Tersan; Tuads; Tulisan; Agrichem Flowable Thiram; Fernide; FS Thiram 15% Dust; Pomarsol®; Spotret 75 WDG; Hortag Thiram. A thiram-based foliage fungicide. Fungicide with animal repellent properties. Also used in vulcanization and as a bacteriostat. mp = 146-148°; insoluble in H_2O, soluble in non-polar organic solvents; LD_{50} (rat orl) = 650 mg/kg. *Universal Crop Protection Ltd.; Agrichem (International) Ltd.; ICI Chem & Polymers Ltd.; Ford Smith & Co Ltd.; Bayer AG; W A Cleary; Avon Packers Ltd.*

489   Tolclofos-methyl
57018-04-9   260-515-3
$C_9H_{11}Cl_2O_2PS$
O-(2,6-Dichloro-4-methylphenyl)   O,O-dimethyl phosphorothioate.
Basilex; Rizolex; S-3349; Dichloro-4-methyl-phenyl-O,O-dimethyl phosphorothioate. Wettable powder containing tolclofos-methyl; protective fungicide for use on all ornamentals and some edible crops against *Rhizoctonia*. mp = 78-80°. *Fisons plc, Horticultural Div.; Schering Agriculture.*

490   Topas 100
66246-88-6   266-275-6
$C_{13}H_{15}Cl_2N_3$
1-[2-(2,4-Dichlorophenyl)pentyl]-1H-1,2,4-triazole.
Award;   CGA   71818;   Topaz;   Topaze; penconazole. Systemic fungicide with protective and curative action. Inhibits biosynthesis of ergosterol in the cell membrane. Topas 100 is an emulsifiable concentrate containing 100 g/l penconazole; used for control of powdery mildew. mp = 60°; soluble in H_2O (70 mg/l), more soluble in organic solvents; LD_{50} (rat orl) = 2125 mg/kg. *Ciba-Geigy Agrochemicals.*

491   Triadimenol
55219-65-3   9724   259-537-6
$C_{14}H_{18}ClN_3O_2$
β-(4-Chlorophenoxy)-α-(1,1-dimethylethyl)-1,2,4-Triazole-1-ethanol.
Spinnaker; Summit; Baytan; Bayfidan; . Fungicide used to control powdery mildew, rusts, and rhychosporium in winter and spring crops of wheat, barley, oats, and rye.Inhibits ergosterol biosynthesis in fungi. Emulsifiable concentrate containing 250 g/l triadimenol; used to control powdery mildew, rusts and rhychosporium in winter and spring crops of cereals, beet and brassicas. *Shell UK; Bayer AG.*

492   Triforine
26644-46-2   247-872-0
$C_{10}H_{14}Cl_6N_4O_2$
N,N'-[1,4-Piperazinediylbis(2,2,2-trichloroethylidene)]bisformamide.
Fairy Ring Destroyer; Funginex; Ortho Rose Pride Funginex Rose and Shrub Disease Control; Saprol; Denarin; Cela W-524; biformylchlorazin; CA 70203; CA 73021; CELA 50; CW 524; Piperazinediylbis(2,2,2-trichloroethylidine)bis-for-mamide;   Piperazinediylbis(α,α,α-trichloro-ethyl-idene))bis(formamide). Emulsifiable concen-trate containing 190 g/l triforine; a systemic fungicide. mp = 155°; slightly soluble in H_2O (28 mg/l), insoluble in organic solvents. *Synchemicals Ltd.*

493   Triphenyltin Hydroxide
76-87-9   9875   200-990-6
Triphenyltin hydroxide.
Ashlade Flotin; Du-Ter; Du-Ter®; Farmatin; Quadrangle Super-Tin 4L; Super Tin® 4L. Suspension concentrate containing 625 g triphenyltin hydroxide per liter; used for prevention of potato blight and disease control in sugar beet. Flowable fungicide for pecans, potatoes, sugarbeets; restricted use. *Ashlade Formulations Ltd,; ICI Chem. & Polymers Ltd; Chiltern Farm Chemicals Ltd,; Griffin; Farm Protection Ltd; Quadrangle Agrochemicals.*

494   Tubotox
88-85-7   3341   201-861-7
$C_{10}H_{12}N_2O_5$
4,6-Dinitro-2-sec-butylphenol.Dinoseb; DNOSBP; F-ISO; Caldon; Vertac General and Selective Weed Killer; Basanite; Chemox General & PE; Chemsect; Dinitrax; Dinitro-3; Dinitro General; Drexel Dynamite 3; Dynamite; Elgetol 318; Hel-Fire; Kiloseb; Nitropone C; Subitex; Unicrop DNBP; Vertac Dinitro Weed Killer 5; Dynanap; Premerge Plus, with Dinitro; Klean Krop; DNBP; WSX-8365; Chemox PE; Dow General; Premerge; Dinitrobutyl phenol; Dinoseb Phenol; Dinitro; DNPB; Vertac General Weed Killer; butaphene; BNP 30; pnosbp; nitropone; phenotan; sparic; spurge; unicorp dnbp; vertac dinitro weed killer; DNSBP; dow general weed killer; dow selective weed killer; Sinox general. Fungicide based on dinoseb. *May & Baker Ltd. See dinoseb.*

495   Vancide® 89
133-06-2   1815   205-087-0
$C_9H_8Cl_3NO_2S$
2-[(Trichloromethyl)thio]-1H-isoindole-1,3(2H)-dione.
3a,4,7,7a-Tetrahydro-2-[(trichloromethyl)thio]-1H-isoindole-1,3(2H)dione; N-(trichloromethylthio)-4-cyclohexene-1,2-dicaboximide; N-trichloro-methyl-thio-3a,4,7,7a-tetrahydropthalimide; N-tri-chloromethyl-mercapto-4-cyclohexene-1,2 dicar-boximide;   ENT   26538;   N-(trichloro-methylmercapto)-δ^4-tetrahydorphthalimide;   SR-406; Merpan; Orthocide-406; Captan Granular; Hormone Rooting Powder; Orthocide; PP Captan 83; Folpet; Folpan; Phaltan. Captan as dry granules; used for controlling soil borne fungal diseases in tomato, lettuce and strawberry. Fungicide for natural and synthetic rubber compounds containing susceptible plasticizers; industrial preservative for vinyl, polyethylene, paint, lacquer, soap, wallpaper flour paste.

Bacteriostat in soap. mp = 178°; d = 1.74. *R. T. Vanderbilt Co Inc.; Wheatley Chemical Co Ltd.; Murphy Chemical Co Ltd.; Makhteshim Chemical Works Ltd.; Monsanto (Solaris); ICI Agrochemicals.*

496   Vinclozolin

50471-44-8   10122   256-599-6

$C_{12}H_9Cl_2NO_3$

3-(3,5-Dichlorophenyl)-5-ethenyl-5-methyl-2,4-oxazolidinedione.

Mascot Contact Turf Fungicide; Power Drive; Ronilan; Ronilan® DF,FL; vinclozalin; dichlorophenyl-5-ethenyl-5-methyl-2,4-oxazolidinedione; Ornalin; Vorlan; 3-(3,5-dichlorophenyl)-5-ethenyl-5-methyloxazolidinedione. Protectant fungicide for oilseed rape, peas, beans and turf. Contact fungicide for use in vines, fruit, strawberries, vegetables, ornamentals, hops, etc. mp = 108°; $bp_{0.05}$ = 131°; soluble in H_2O (1 g/l), more soluble in organic solvents; LD_{50} (rat orl) = 10 g/kg. *BASF AG; Kommer-Brookwick Ltd.; Rigby Taylor Ltd.*

497   Zinc

7440-66-6   10255   231-175-3

Zn

Zinc.

Metallic element; Zn; used in alloys, for galvanizing iron and other metals, and in fungicides. mp = 419.5°; bp = 908°; d^{25} = 7.14. *Aldrich; Cerac; Cuproquim; Ferro/Bedford; Pasminco Europe; Zinc Corp. of Am.*

498   Zineb

12122-67-7   10300   235-180-1

$(C_4H_6N_2S_4Zn)_x$

[[1,2-Ethanediylbis[carbamodithioato]](2-)]zinc. zinc ethylenebis(dithiocarbamate) (polymeric); zinebe, Zidan; Aaphytora; Acuprex; Aspor; Dipher; Dithane Z 78; Diiner; Ditiozin; Enozin; Hexaphane; Kypzin; Lonacol; Parzate; Permilan; Phytox; Polyram-Z; Sepineb; Tiezene; Tritoftorol; Zinosan; Zinugec; Zidanit. A fungicide and insecticide. Foliar fungicide with protective action, insecticide. Repellent to birds and rodents. Used for control of fungi in fruits, vines, vegetables and ornamentals. Controls scab in apples and pears. dec 157°; soluble in H_2O 10 mg/l, insoluble in organic solvents; LD_{50} (rat orl) > 5200 mg/kg. *Agrimont; Diachem; Bayer; Pennwalt Holland; Makhteshim Chemical Works Ltd; Rhône Poulenc, Visplant; Rohm & Haas.*

Fungicides containing Benomyl

499   Polycote Pedigree

Benomyl + iodofenphos + metalaxyl; a fungicide and insecticide seed coating for seeds.   *Seedcote*

Systems Ltd. See benomyl, iodofenphos, metalaxyl.

Fungicides containing Captafol

500   Milcap

Fungicide containing ethirimol and captafol for use on wheat. *ICI Chem & Polymers Ltd.* See ethirimol, captafol.

Fungicides containing Captan

501   Aliette Extra

Captan + fosetyl + aluminum + thiabendazole. Fungicide seed dressing for peas. *Embetec Crop Protection Ltd.* See captan, fosetyl-aluminum, thiabendazole.

502   Gammalex

Captan + lindane.

Captan.2 γ-HCH; insecticide and fungicide seed dressing for brassicas and oilseed rape. *ICI Agrochemicals; ICI Chem. & Polymers Ltd.* See captan, lindane.

503   Kapitol

Captan + nuarimol; systemic fungicide for apple and pear trees. *DowElanco Ltd.* See captan; nuarimol.

504   Topas C 50WP

Captan + penconazole; protectant fungicide for apple and pear trees. *Ciba-Geigy Agrochemicals.*

Fungicides containing Carbendazim

505   Ashlade Cosmic FL

Suspension concentrate containing 40 g carbendazim, 320 g maneb and 90 g tridemorph per liter. systemic fungicide for cereals. Ashlade Formulations Ltd. See carbendazim, maneb, tridemorph.

506   Bayleton® BM

A wettable powder systemic fungicide containing 12.5% w/w triadimenol and 25% w/w carbendazim; used to control eyespot, mildew and early attacks of yellow and brown rust on winter wheat and winter barley, rhynchosporium on winter barley and eyespot and mildew on winter rye. Bayer AG. See triadimenol, carbendazim.

507   Bravocarb

Liquid formulation containing 100 g carbendazim and 450 g chlorothalonil per liter as a suspension concentrate; systemic fungicide. *Fermenta ASC Europe Ltd.* See carbendazim, chlorothalonil.

508 Campbell's MC Flowable
A suspension concentrate containing 62 g carbendazim [10605-21-7] and 400 g maneb [12427-38-2] per liter. systemic fungicide for cereals. *MTM AgroChemicals Ltd. See* carbendazim, maneb.

509 Corbel® Duo
Fenpropimorph + Carbendazim.
Combination of Fenpropimorph and Carbendazim; used as a fungicide. *BASF AG. See* fenpropimorph, carbendazim.

510 Delsene® M Flowable
83601-81-4, 12427-38-2
A suspension concentrate containing 50 g carbendazim and 320 g maneb per liter; systemic fungicide for cereals. *DuPont UK. See* carbendazim, maneb.

511 Early Impact
Carbendazim + flusilazole.
A suspension concentrate containing 150 g carbendazim and 94 g flusilazole per liter; systemic fungicide for use on cereals. *DuPont UK. See* carbendazim, flusilazole.

512 Headland Dual
Carbendazim + maneb.
A suspension concentrate containing 62 g carbendazim and 400 g maneb per liter; systemic fungicide for use with cereals. *WBC Technology Ltd. See* carbendazim, maneb.

513 Hickstor 6 .2 MBC
Carbendazim + tecnazene.
Carbendazim and tecnazene; Protectant fungicide and sprout suppressant for stored potatoes. *Hickson & Welch Ltd. See* carbendazim, tecnazene.

514 Hispor 45WP
Carbendazim + maneb.
Carbendazim and maneb; systemic fungicide for winter cereals. *Ciba Geigy Agrochemicals. See* carbendazim, maneb.

515 Hortag Carbotec
Carbendazim, tecnazene.
Carbendazim and tecnazene; protectant fungicide and sprout suppressant for stored potatoes. *Avon Packers Ltd. See* carbendazim, tecnazene.

516 Hortag Tecnacarb Dust
Carbendazim + tecnazene.
Carbendazim and tecnazene; protectant fungicide and sprout suppressant for stored potatoes. *Avon Packers Ltd. See* carbendazim, tecnazene.

517 Kombat
Carbendazim + mancozeb.
Systemic fungicide for cereals. *Hoechst UK. See* carbendazim and mancozeb.

518 Mancarb Plus
A liquid formulation containing 80 g carbendazim, 150 g chlorothalonil and 200 g maneb per liter as a suspension concentrate; eradicant fungicide for use on cereals. *Ashlade Formulations Ltd. See* carbendazim, chlorothalonil and maneb.

519 MaxiMate
A suspension concentrate containing 62 g carbendazim and 400 g maneb per liter; systemic fungicide for cereals. *Farmers Crop Chemicals Ltd. See* carbendazim, maneb.

520 Multi-W FL
A suspension concentrate containing 50 g carbendazim [83601-81-4] and 320 g maneb [12427-38-2] per liter; systemic fungicide for cereals. *Pan Britannica Industries Ltd. See* carbendazim, maneb.

521 Punch® C
Eradicant fungicide for use on cereal crops. *DuPont UK. See* carbendazim, flusilazole.

522 Ridomil MBC 60WP
Carbendazim + metalaxyl; a fungicide mixture used to prevent the spread of fruit and vegetable storage diseases. *Ciba-Geigy Agrochemicals. See* carbendazin, metalaxyl.

523 Sportak Alpha
A suspension concentrate containing 266 g prochloraz [67747-09-5] and 100 g carbendazim [83601-81-4] per liter; systemic fungicide for cereals. *Schering Agrochemicals Ltd.*

524 Squadron
A suspension concentrate containing 100 g carbendazim [83601-81-4] and 275 g maneb [12427-38-8] per liter; systemic fungicide for cereals. *Quadrangle Agrochemicals.*

525 Tripart® Legion
Carbendazim [10605-21-7] 50 g and maneb [12427-38-2] 320 g/liter mixture; for use as a fungicide in cereals. *Tripart Farm Chemicals Ltd.*

526 Tripart® Victor
A liquid formulation containing 80 g carbendazim, 150 g chlorothalonil and 200 g maneb per liter as a suspension concentrate; eradicant fungicide for use on cereals. *Tripart Farm Chemicals Ltd.*

Fungicides containing Carboxin

527 Cerevax, Cerevax Extra

A flowable concentrate containing 360 g carboxin and 20 g thiabendazole; fungicide seed dressing for rye and wheat. Cerevax Extra also contains 25 g imazalil. *ICI AgroChemicals; ICI Chemical & Polymers Ltd.* See carboxin, imazalil, thiabendazole.

528 Vitavax RS Flowable

Carboxin + γ-HCH + thiram; a fungicide and insecticide dressing for oilseed rape. *Uniroyal Chemical Ltd.* See carboxin, γ-HCH, thiram.

Fungicides containing Chlorothalonil

529 BAS 438

Suspension concentrate containing 250g chlorothalonil and 187g fenpropimorph per liter; a systemic fungicide for winter wheat. *BASF plc.* See chlorothalonil, fenpropimorph.

530 Bravocarb

Liquid formulation containing 100 g carbendazim and 450 g chlorothalonil per liter as a suspension concentrate; systemic fungicide. *Fermenta ASC Europe Ltd.* See carbendazim, chlorothalonil.

531 Corbel® CL, Corbel® Star

Chlorothalonil + fenpropimorph.
Suspension concentrate containing 250 g chlorothalonil and 187 g fenpropimorph per liter; a systemic fungicide for winter wheat. *BASF plc.* See chlorothalonil, fenpropimorph.

532 Folio 575FW

Chlorothalonil + metalaxyl.
Suspension concentrate containing 500 g chlorothalonil and 75 g metalaxyl per liter, a systemic fungicide for field crops. *Ciba-Geigy Agrochemicals.* See chlorothalonil, metalaxyl.

533 Impact Excel

Chlorothalonil + flutriafol.
Suspension concentrate containing 300 g chlorothalonil [1897-45-6] and 47 g flutriafol [76674-21-0] per liter; a systemic fungicide. *ICI Agrochemicals.* See chlorothalonil, flutriafol.

534 Lindex-Plus

Flowable concentrate of 43 g fenpropimorph, 545 g γ-HCH and 73 g thiram per liter; combined insecticide and fungicide seed treatment. *DowElanco Ltd.*

535 Mistral CT

A suspension concentrate containing 250 g chlorothalonil and 187 g fenpropimorph per liter; a systemic fungicide for winter wheat. *Rhône-Poulenc Crop Protection Ltd.* See chlorothalonil, fenpropimorph.

536 Sambarin

Suspension concentrate containing 500 g chlorothalonil and 250 g propiconazole per liter; a systemic fungicide for winter wheat. *Ciba-Geigy Agrochemicals.* See chlorothalonil, propiconazole.

Fungicides containing Copper

537 Ashlade SMC

Copper Oxychloride, maneb and sulfur; a fungicide for wheat and barley which also stimulates yields. *Ashlade Formulations Ltd.*

538 Comac Bordeaux Plus

Copper sulfate/lime complex; a protectant fungicide for the control of blight and canker. *McKechnie Chemicals Ltd.*

539 Copper Antracol®

$C_5H_8CuN_2S_4$
[[(1-Methyl-1,2-ethendiyl)bis-[carbamodithioato]](2-)copper.
Combination product for prevention of fungal infections in fruit, grapes, vegetables and arable crops. *Bayer AG.*

540 Cupertine Folpet

Folpet + cupric sulfate.
Bordeaux mixture 80%, folpet 10%; wettable powder used as protective fungicide for foliage application to ornamental and crop plants. *Industrias Quimicas Del Valles SA.* See folpet, cupric sulfate.

541 FS Bordeaux Powder

Copper sulfate/lime complex; a protectant fungicide for the control of blight and canker. *Ford Smith & Co Ltd.*

542 Fungex

A copper-containing fungicide. *Murphy Chemical Co Ltd.*

543 Jeunite

Copper-based fungicide. *Murphy Chemical Co Ltd.*

544 K-Cop

Copper-ammonium complex; agricultural fungicide. *Griffin.*

545 Kocidc® 20/20

Copper hydroxide plus nutritional zinc; wettable powder, agricultural fungicide, bactericide, and nutritional. *Griffin.* See copper hydroxide and nutritional zinc.

546   Macuprax
Bordeaux/cufraneb.
Fungicide composed of basic cupric sulfate and cufraneb. *McKechnie Chemicals Ltd.* See basic cupric sulfate, cufraneb.

547   Monterey Liqui-Cop
Copper + Count + N.
A liquid copper formulation used as a dormant copper spray fungicide on stone fruits and citrus; used as a replacement for Bordeaux mixtures. *Lawn & Garden Products Inc.*

548   Parasiticine
A fungicide containing 57% copper sulfate, sodium carbonate and sodium bicarbonate.

549   Perecot
A copper containing fungicide. *ICI Chem. & Polymers Ltd.*

550   Pereman
A copper fungicide. Plant Protection.

551   Peronoid
A mixture of copper sulfate and lime; a fungicide. *See* cupric sulfate, calcium oxide.

552   Potash Bordeaux Mixture
Contains 6 lb copper sulfate, 2 lb potassium hydroxide, and 50 gallons water. A fungicide.

553   Quindex
Fungicides containing copper 8-quinolinolate; for preservation of textiles, cordage, paper, adhesives and caulking compounds. *Hüls Am.* See copper 8-quinolinolate.

554   Rearguard
Suspension concentrate containing 64% w/v sulfur, 16% w/v maneb, 1% w/v copper oxychloride; for use as an agricultural fungicide. *Universal Crop Protection Ltd.*

555   Ridomil Plus 50WP
Copper oxychloride + metalaxyl; a systemic fungicide mixture used to prevent mildew and root rot in a range of fruit and vegetables. *Ciba-Geigy Agrochemicals.*

556   Segetan
A silver cyanide with a copper complex; used as a seed preservative.

557   Sulfatine
A mixture of 73% sulfur, 20% lime, and 7% copper sulfate; a fungicide used against black rot.

558   Tanalith
Copper/chrome/arsenate waterborne wood preservative to prevent fungal decay and insect attack; for pressure treated timber for construction, fencing, agriculture and any application where timber requires protection. *Hickson & Welch Ltd.*

559   Top-Cop
Mixture of copper sulfate and sulfur; a contact fungicide to control mildew. *Stoller Chemicals Ltd.* See copper sulfate, sulfur.

560   Tripart® Senator Flowable
Copper oxychloride + maneb + sulfur; a fungicide for wheat and barley which also stimulates yield. *Tripart Farm Chemicals Ltd.* See copper oxychloride, maneb, sulfur.

Fungicides containing Cymoxanil

561   Cupertine Super
Bordeaux mixture 90%, cymoxanil 3%; wettable powder used as protective and curative fungicide for foliage application to ornamental and crop plants. *Industrias Quimicas Del Valles SA.* See cupric sulfate, cymoxanil.

562   Curzate® M
Mixture of cymoxanil and mancozeb; used to control potato blight. *DuPont UK.* See cymoxanil, mancozeb.

563   Fytospore
Cymoxanil + mancozeb.
Mixture of cymoxanil and mancozeb; fungicide for the control of potato blight. *ICI Chem & Polymers Ltd.* See cymoxanil, mancozeb.

564   Vironex
Folpet 40%, cymoxanil 4%; wettable powder used as protective fungicide for foliage application to ornamental and crop plants. *Industrias Quimicas Del Vailes SA.* See folpet, cymoxanil.

Fungicides containing Dithiocarbonate

565   Merolan
Fungicide containing dithiocarbonate. *ICI Chem & Polymers Ltd.*

Fungicides containing Ethirimol

566   Ferrax
Ethirimol + flutriafol + thiabendazole.
Flowable concentrate containing 400g ethirimol, 30g flutriafol and 10g thiabendazole per liter; fungicide seed treatment for barley. *Bayer AG; Bayer plc; Dow Elanco Ltd.* See ethirimol, flutriafol, thiabendazole.

567   Milcap

Fungicide containing ethirimol and captafol for use on wheat. *ICI Chem & Polymers Ltd. See* ethirimol, captafol.

Fungicides containing Fenpropimorph

568   BAS 438

Suspension concentrate containing 250g chlorothalonil and 187g fenpropimorph per liter; a systemic fungicide for winter wheat. *BASF plc. See* chlorothalonil, fenpropimorph.

569   Corbel® Duo

Fenpropimorph + Carbendazim.
Combination of Fenpropimorph and Carbendazim; used as a fungicide. *BASF AG. See* fenpropimorph, carbendazim.

570   Lindex-Plus

Flowable concentrate of 43 g fenpropimorph, 545 g γ-HCH and 73 g thiram per liter; combined insecticide and fungicide seed treatment. *DowElanco Ltd.*

571   Mistral CT

A suspension concentrate containing 250 g chlorothalonil and 187 g fenpropimorph per liter; a systemic fungicide for winter wheat. *Rhône-Poulenc Crop Protection Ltd. See* chlorothalonil, fenpropimorph.

Fungicides containing Fentin

572   Brestan

Mixture of fentin acetate and maneb; used for control of potato blight. *Hoechst UK. See* fentin acetate, maneb.

573   Endspray

Fentin hydroxide + metoxuron.
Mixture of fentin hydroxide and metoxuron; used for control of potato blight. *Pan Britannica Industries Ltd. See* fentin hydroxide, metoxuron.

574   Hytin

Fentin acetate + maneb.
Mixture of fentin acetate and maneb; used for control of potato blight. *Agrichem (International) Ltd. See* fentin acetate, maneb.

575   Tubotin

Fungicide with fentin hydroxide as the active component. *May & Baker Ltd. See* fentin hydroxide.

Fungicides containing Flusilazole

576   Early Impact

Carbendazim + flusilazole.
A suspension concentrate containing 150 g carbendazim and 94 g flusilazole per liter; systemic fungicide for use on cereals. *DuPont UK. See* carbendazim, flusilazole.

Fungicides containing Flutriafol

577   Ferrax

Ethirimol + flutriafol + thiabendazole.
Flowable concentrate containing 400g ethirimol, 30g flutriafol and 10g thiabendazole per liter; fungicide seed treatment for barley. *Bayer AG; Bayer plc; Dow Elanco Ltd. See* ethirimol, flutriafol, thiabendazole.

578   Impact Excel

Chlorothalonil + flutriafol.
Suspension concentrate containing 300 g chlorothalonil [1897-45-6] and 47 g flutriafol [76674-21-0] per liter; a systemic fungicide. *ICI Agrochemicals. See* chlorothalonil, flutriafol.

Fungicides containing Folpet

579   Cupertine Folpet

Folpet + cupric sulfate.
Bordeaux mixture 80%, folpet 10%; wettable powder used as protective fungicide for foliage application to ornamental and crop plants. *Industrias Quimicas Del Valles SA. See* folpet, cupric sulfate.

580   Vironex

Folpet 40%, cymoxanil 4%; wettable powder used as protective fungicide for foliage application to ornamental and crop plants. *Industrias Quimicas Del Vailes SA. See* folpet, cymoxanil.

Fungicides containing Fosetyl-aluminum

581   Aliette Extra

Captan + fosetyl + aluminum + thiabendazole.
Fungicide seed dressing for peas. *Embetec Crop Protection Ltd. See* captan, fosetyl-aluminum, thiabendazole.

Fungicides containing Fuberidazole

582   Baytan

Powder mixture of fuberidazole and triadimenol; seed dressing for cereals. *ICI Agrochemicals. See* Baytan®, fuberidazole, triadimenol.

583   Baytan® IM

Dry powder mixture of fuberidazole, imazalil and triadimenol; seed dressing for barley. *Bayer plc.*

584   Neo-Voronit®

Fuberidazole, sodium-N-dimethyldithiocarbamate; Nonmercurial liquid seed dressing for treatment of cereal seed against smuts and snow mold. *Bayer AG. See* fuberidazole, sodium-N-dimethyldithiocarbamate.

585   Sibutol®

Bitertanol [55179-31-2] and fuberidazole [3878-19-1]; seed dressing for cereals for control of loose smuts, stinking smuts, flag smuts; especially effective against dwarf bunt of wheat and seed-borne snow mold. *Bayer AG. See* bitertanol, fuberidazole.

Fungicides containing Guazatine

586   Rappor Plus

A fungicide seed dressing for barley and oats. *DowElanco Ltd. See* guazatine, imazalil.

Fungicides containing Iprodione

587   Compass

Mixture of iprodione and thiophanate-methyl; systemic fungicide for field corps. *Rhône-Poulenc Crop Protection Ltd. See* iprodione, thiophanate-methyl.

588   Polycote Prime

Powder mixture of iprodione, metalaxyl and thiabendazole; a polymer seed coating for carrots. *Seedcote Systems Ltd. See* iprodione, metalaxyl, thiabendazole.

Fungicides containing Magnesium

589   Sulfarine

A mixture of magnesium sulfate with 15% sulfuric acid; used against potato scab.

Fungicides containing Mancozeb

590   Curzate® M

Mixture of cymoxanil and mancozeb for potato blight. *DuPont UK. See* cymoxanil, mancozeb.

591   Fytospore

Cymoxanil + mancozeb.
Mixture of cymoxanil and mancozeb; fungicide for the control of potato blight. *ICI Chem & Polymers Ltd. See* cymoxanil, mancozeb.

592   Kombat

Carbendazim + mancozeb.
Systemic fungicide for cereals. *Hoechst UK. See* carbendazim and mancozeb.

593   Recoil

A wettable powder containing 10% w/w oxadixyl and 56% w/w mancozeb; fungicide used to control foliar and tuber blight in potatoes. *Schering Agrochemicals Ltd. See* oxadixyl, mancozeb.

595   Systol M

Mixture of cymoxanil and mancozeb; used to control potato blight. *Quadrangle Agrochemicals. See* cymoxanil, mancozeb.

596   Trustan®

Mixture of cymoxanil and mancozeb; systemic fungicide to control potato blight. *DuPont UK. See* cymoxanil, mancozeb.

Fungicides containing Maneb

597   Amazin®

Maneb with zineb; fungicide. *Aceto. See* maneb, zineb.

598   Ashlade Blight Fungicide

Mixture of cymoxanil and mancozeb; used to control potato blight. *Ashlade Formulations Ltd. See* cymoxanil, mancozeb.

599   Ashlade Cosmic FL

A suspension concentrate containing 40 g carbendazim, 320 g maneb and 90 g tridemorph per liter. systemic fungicide for cereals. *Ashlade Formulations Ltd. See* carbendazim, maneb, tridemorph.

600   Ashlade SMC

Copper Oxychloride, maneb and sulfur; a fungicide for wheat and barley which also stimulates yields. *Ashlade Formulations Ltd. See* copper oxychloride, maneb, sulfur.

601   Bravocarb

A liquid formulation containing 100 g carbendazim and 450 g chlorothalonil per liter as a suspension concentrate; systemic fungicide. *Fermenta ASC Europe Ltd. See* carbendazim, chlorothalonil.

602   Brestan

Mixture of fentin acetate and maneb; used for control of potato blight. *Hoechst UK. See* fentin acetate, maneb.

603   Campbell's MC Flowable

A suspension concentrate containing 62 g carbendazim [10605-21-7] and 400 g maneb [12427-38-2] per liter. systemic fungicide for cereals. *MTM AgroChemicals Ltd. See* carbendazim, maneb.

604   Cupertine

Maneb + cupric sulfate.

Bordeaux mixture 84%, maneb 8%; wettable powder used as protective fungicide for foliage application to ornamental and crop plants. *Industrias Quimicas Del Valles SA. See* maneb, cupric sulfate.

605   Delsene® M Flowable

83601-81-4, 12427-38-2

A suspension concentrate containing 50 g carbendazim and 320 g maneb per liter; systemic fungicide for cereals. *DuPont UK. See* carbendazim, maneb.

606   Headland Dual

Carbendazim + maneb.

A suspension concentrate containing 62 g carbendazim and 400 g maneb per liter; systemic fungicide for use with cereals. *WBC Technology Ltd. See* carbendazim, maneb.

607   Hispor 45WP

Carbendazim + maneb.

Carbendazim and maneb; systemic fungicide for winter cereals. *Ciba Geigy Agrochemicals. See* carbendazim, maneb.

608   Hytin

Fentin acetate + maneb.

Mixture of fentin acetate and maneb; used for control of potato blight. *Agrichem (International) Ltd. See* fentin acetate, maneb.

609   Mancopper

An ethylene bisdithiocarbamate-mixed metal complex containing about 13.7% manganese and about 4% copper; used as a fungicide and pesticide.

610   Manex

Suspension concentrate containing maneb and zinc; a protectant fungicide against potato blight, mildew and rust. *Chiltern Farm Chemicals Ltd; Griffin; Quadrangle Agrochemicals; L W Vass (Agricultural) Ltd. See* maneb.

611   Mazin®

Wettable powder containing maneb 80% w/w and zinc oxide; protective fungicide against potato blight. *Universal Crop Protection Ltd. See* maneb, zinc oxide.

612   Multi-W FL

A suspension concentrate containing 50 g carbendazim [83601-81-4] and 320 g maneb [12427-38-2] per liter; systemic fungicide for cereals. *Pan Britannica Industries Ltd. See* carbendazim, maneb.

613   Patafol Plus

Manganese zinc ethylenebisdithiocarbamate (maneb, zineb) and ofurace (Patafol) mixture; for control of potato blight. (Sold in UK for Chevron Chemical Co.) . *ICI Chem & Polymers Ltd. See* maneb, zineb, patafol.

614   Pro-Tex

Maneb (32.63%) and triphenyltin hydroxide (4.72%) solution; flowable fungicide for potatoes and sugar beets; restricted use. *Griffin. See* maneb, triphenyltin hydroxide.

615   Rearguard

Suspension concentrate containing 64% w/v sulfur, 16% w/v maneb, 1% w/v copper oxychloride; for use as an agricultural fungicide. *Universal Crop Protection Ltd. See* sulfur, maneb, copper oxychloride.

616   Squadron

A suspension concentrate containing 100 g carbendazim [83601-81-4] and 275 g maneb [12427-38-8] per liter; systemic fungicide for cereals. *Quadrangle Agrochemicals. See* carbendazim, maneb.

617   Tinman

Mixture of fentin hydroxide, maneb, and zineb; used for control of potato blight. *Chiltern Farm Chemicals Ltd. See* fentin hydroxide, maneb, zineb.

618   Trimanzone

Mixture of ferbam, maneb and zineb; a fungicide. *Pennwalt Chemicals Ltd. See* ferbam, maneb, zineb.

619   Trimastan

Mixture of fentin acetate and maneb; used for control of potato blight. *Pennwalt Chemicals Ltd. See* fentin acetate, maneb.

620   Tripart® Legion

Carbendazim [10605-21-7] 50 g and maneb [12427-38-2] 320 g/liter mixture; for use as a fungicide in cereals. *Tripart Farm Chemicals Ltd. See* carbendazim, maneb.

621   Tripart® Senator Flowable

Copper oxychloride + maneb + sulfur; a fungicide for wheat and barley which also stimulates yield. *Tripart Farm Chemicals Ltd.*

Fungicides containing Mercury

623 Abavit B
Organo-mercury seed dressing. *Murphy Chemical Co Ltd.*

624 Abavit S
Organo-mercurial dip. *Murphy Chemical Co Ltd.*

625 Agrosol
Liquid mercury seed dressing, a fungicide. *Plant Protection.*

626 Aretan
A mercury-based fungicide. *ICI Chem.*

627 Aspulum
Mercury derivative of chlorophenol; seed preservative.

628 Fungitex 656
A proprietary trade name for a solution of an organo-mercuric complex; used as a durable mildew proofing agent for textiles. *Ciba plc.*

629 Fusariol
A mercury-formaldehyde preparation. A seed preservative.

630 Mergamma
Mixture of γ-HCH (lindane) and phenylmercury acetate; fungicide seed dressing for cereals. *ICI Chem. See* lindane, phenylmercury acetate.

631 Murcurite
A mercury-containing fungicide. *Murphy Chemical Co Ltd.*

632 Murfixtan
A mercury fungicide. *Murphy Chemical Co Ltd.*

633 Phemox
Mercury containing fungicide. *Murphy Chemical Co Ltd.*

634 PMAS
Mercurial fungicide for prevention and control of pink and gray snow mold. *W A Cleary. See* phenylmercuric acetate.

635 Presol W
Mercury fungicide solution. *Great Lakes Europe.*

636 Sublimoform
A mercury-formaldehyde preparation; used as an antifungal seed preservative.

637 Uspulun
Hydroxymercurichlorophenol sulfates.
A material containing sodium sulfate, sodium hydroxide, aniline, and mercury-chloro-phenol; used as a fungicide. Seed dressing for control of fungal diseases on cereals, rice, cotton, and vegetables. *Bayer AG.*

Fungicides containing Metalaxyl

638 Folio 575FW
Chlorothalonil + metalaxyl.
Suspension concentrate containing 500 g chlorothalonil and 75 g metalaxyl per liter, a systemic fungicide for field crops. *Ciba-Geigy Agrochemicals. See* chlorothalonil, metalaxyl.

639 Polycote Pedigree
Benomyl + iodofenphos + metalaxyl; a fungicide and insecticide seed coating for seeds. *Seedcote Systems Ltd. See* benomyl, iodofenphos, metalaxyl.

640 Polycote Prime
Powder mixture of iprodione, metalaxyl and thiabendazole; a polymer seed coating for carrots. *Seedcote Systems Ltd. See* iprodione, metalaxyl, thiabendazole.

641 Ridomil MBC 60WP
Carbendazim + metalaxyl; a fungicide mixture used to prevent the spread of fruit and vegetable storage diseases. *Ciba-Geigy Agrochemicals. See* carbendazin, metalaxyl.

642 Ridomil Plus 50WP
Copper oxychloride + metalaxyl; Systemic fungicide mixture to prevent mildew and root rot in a range of fruit and vegetables. *Ciba-Geigy Agrochemicals. See* copper oxychloride, metalaxyl.

Fungicides containing Metiram

643 Avisol® G
Metiram-cymoxanil fungicide for potatoes, vines, and other crops. *BASF AG. See* metiram, cymoxanil.

644 Pallinal®
Metiram, nitrothal-isopropyl; for control of powdery mildew in fruit, vegetables, hops, ornamentals. *BASF AG.*

645 Pallitop®
Nitrothal-isopropyl, metiram; for control of powdery mildew in apples, pears. *BASF AG; BASF plc.*

Fungicides containing Nuarimol

646 Kapitol
Captan + nuarimol; systemic fungicide for apple and pear trees. *DowElanco Ltd.*

Fungicides containing Oruface

647 Patafol Plus
Manganese zinc ethylenebisdithiocarbamate (maneb, zineb) and ofurace (Patafol) mixture; for control of potato blight. *(Sold in UK for Chevron Chemical Co.). ICI Chem & Polymers Ltd. See* maneb, zineb, patafol.

Fungicides containing Oxadixyl

648 Recoil
Wettable powder containing 10% w/w oxadixyl and 56% w/w mancozeb; fungicide for control of foliar and tuber blight in potatoes. *Schering Agrochemicals Ltd. See* oxadixyl, mancozeb.

Fungicides containing Penconazole

650 Topas C 50WP
Captan + penconazole; protectant fungicide for apple and pear trees. *Ciba-Geigy Agrochemicals.*

Fungicides containing Prochloraz

651 Sportak Alpha
A suspension concentrate containing 266 g prochloraz [67747-09-5] and 100 g carbendazim [83601-81-4] per liter; systemic fungicide for cereals. *Schering Agrochemicals Ltd.*

Fungicides containing Propiconazole

652 Sambarin
Suspension concentrate containing 500 g chlorothalonil and 250 g propiconazole per liter; a systemic fungicide for winter wheat. *Ciba-Geigy Agrochemicals. See* chlorothalonil, propiconazole.

Fungicides containing Sorbic Acid

653 Aflaban
Feed preservative based on sorbic acid; growth inhibitor for molds, yeast and bacteria in animal feeds. *Monsanto Co.*

654 Brimstone Plus
Mixture of potassium sorbate, sodium metabisulfate and sodium propionate; a broad-spectrum fungicide for field crops. *Mandops (UK) Ltd. See* potassium sorbate, sodium metabisulfate, sodium propionate.

Fungicides containing Sulfur

655 Ashlade SMC
Copper Oxychloride, maneb and sulfur; a fungicide for wheat and barley which also stimulates yields. *Ashlade Formulations Ltd. See* copper oxychloride, maneb, sulfur.

656 Fernasul
A lime and sulfur fungicide. *Plant Protection.*

657 Rearguard
Suspension concentrate containing 64% w/v sulfur, 16% w/v maneb, 1% w/v copper oxychloride; for use as an agricultural fungicide. *Universal Crop Protection Ltd. See* sulfur, maneb, copper oxychloride.

658 S-oils
Sulfur-containing oils obtained by the distillation of crude petroleum oil in the presence of sulfur. They have a strong antiseptic action against wood-destroying fungi.

659 Sulfatine
A mixture of 73% sulfur, 20% lime, and 7% copper sulfate; a fungicide used against black rot.

660 Sulsol
A proprietary trade name for a colloidal sulfur preparation for horticultural purposes.

661 Top-Cop
Mixture of copper sulfate and sulfur; a contact fungicide to control mildew. *Stoller Chemicals Ltd. See* copper sulfate, sulfur.

662 Tripart® Senator Flowable
Copper oxychloride + maneb + sulfur; a fungicide for wheat and barley which also stimulates yield. *Tripart Farm Chemicals Ltd. See* copper oxychloride, maneb, sulfur.

Fungicides containing Tecnazene

663 Hickstor 6 .2 MBC
Carbendazim + tecnazene.
Carbendazim and tecnazene; Protectant fungicide and sprout suppressant for stored potatoes. *Hickson & Welch Ltd. See* carbendazim, tecnazene.

664 Hortag Carbotec
Carbendazim, tecnazene.
Carbendazim and tecnazene; protectant fungicide and sprout suppressant for stored potatoes. *Avon Packers Ltd. See* carbendazim, tecnazene.

665   Hortag Tecnacarb Dust
Carbendazim + tecnazene.
Carbendazim and tecnazene; protectant fungicide
and sprout suppressant for stored potatoes. *Avon
Packers Ltd.* See carbendazim, tecnazene.

666   Hortag Tecnazene Plus
Tecnazene + thiabendazole.
Dustable powder containing 6% w/w tecnazene
[117-18-0] and 1.8% thiabendazole [148-79-8]; a
protectant fungicide and potato sprout
suppressant. *Avon Packers Ltd.* See tecnazene,
thiabendazole.

667   Hytec Super
Tecnazene + thiabendazole.
Dustable powder containing 6% w/w tecnazene
[117-18-0] and 1.8% thiabendazole [148-79-8]; a
protectant fungicide and potato sprout
suppressant. *Agrichem (International) Ltd.* See
tecnazene, thiabendazole.

668   Tripart® Arena 6 +xg TBZ
Dustable powder containing 6% (w/w) tecnazene
[117-18-0] and 1.8% thiabendazole [148-79-8]; a
protectant fungicide and potato sprout
suppressant. *Tripart Farm Chemicals Ltd.* See
tecnazene, thiabendazole.

669   Tripart® Arena Plus
6% (w/w) tecnazene [117-18-0] and 2%
carbendazim [83601-81-4]; protectant fungicide
and sprout suppressant for stored potatoes. *Tripart
Farm Chemicals Ltd.* See tecnazene, carbendazim.

Fungicides containing Thiabendazole

670   Aliette Extra
Captan + fosetyl-aluminum + thiabendazole.
Fungicide seed dressing for peas. *Embetec Crop
Protection Ltd.*

671   Apron Combi
Mixture of metalaxyl, thiram and thiabendazole; a
protectant fungicide for pea and bean seeds. *Ciba-
Geigy Agrochemicals.* See metalaxyl, thiram,
thiabendazole.

672   Apron T
Mixture of metalaxyl and thiabendazole;
protectant fungicide for grass seed. *Ciba-Geigy
Agrochemicals.* See metalaxyl, thiabendazole.

673   Ascot
Mixture of thiabendazole and thiram; fungicide
seed dressing. *Ciba-Geigy Agrochemicals.* See
thiabendazole, thiram.

674   Cerevax, Cerevax Extra
A flowable concentrate containing 360 g carboxin
and 20 g thiabendazole; fungicide seed dressing
for rye and wheat. Cerevax Extra also contains 25
g imazalil. *ICI AgroChemicals; ICI Chemical &
Polymers Ltd.* See carboxin, imazalil,
thiabendazole.

675   Ferrax
Ethirimol + flutriafol + thiabendazole.
Flowable concentrate containing 400g ethirimol,
30g flutriafol and 10g thiabendazole per liter;
fungicide seed treatment for barley. *Bayer AG;
Bayer plc; Dow Elanco Ltd.* See ethirimol,
flutriafol, thiabendazole.

676   Hy-TL
Thiabendazole + thiram.
Mixture of thiabendazole [148-79-8] and thiram
[137-26-8]; fungicide seed dressing. *Agrichem
(International) Ltd.* See thiabendazole, thiram.

677   Hy-Vic
Thiabendazole + thiram.
Mixture of thiabendazole [148-79-8] and thiram
[137-26-8]; fungicide seed dressing. *Agrichem
(International) Ltd.* See thiabendazole, thiram.

678   Polycote Prime
Powder mixture of iprodione, metalaxyl and
thiabendazole; a polymer seed coating for carrots.
Seedcote Systems Ltd.

679   Seedtect
Mixture of imazalil and thiabendazole; fungicide
treatment of potatoes at planting time. *MSD Agvet.*
See imazalil, thiabendazole.

680   Storaid Dust, Storite SS
Dustable powder containing 6% w/w tecnazene
[117-18-0] and 1.8% w/w thiabendazole [148-79-
8]; a protectant fungicide and potato sprout
suppressant. *MSD Agvet.* See tecnazene,
thiabendazole.

681   Tubazole
Mixture containing nonylphenoxypoly
(ethyleneoxy)ethanol-iodine complex and
thiabendazole; used for controlling various
diseases in stored potatoes. *Dean Agrochemicals
Ltd.* See thiabendazole.

Fungicides containing Thiophanate-methyl

682   Compass
Mixture of iprodione and thiophanate-methyl;
systemic fungicide for field corps. *Rhône-Poulenc
Crop Protection Ltd.*

Fungicides containing Thiram

683 Favour

Metalaxyl + thiram.
Mixture of metalaxyl and thiram; protectant fungicide against downy mildew in lettuce. *Ciba-Geigy Agrochemicals*. See metalaxyl, thiram.

684 Hydra-guard

Lindane + thiram.
Mixture of lindane and thiram; seed dressing for brassica crops. *Agrichem (International) Ltd*. See lindane, thiram.

685 Vitavax RS Flowable

Carboxin + γ-HCH + thiram; a fungicide and insecticide dressing for oilseed rape. *Uniroyal Chemical Ltd*. See carboxin, γ-HCH, thiram.

Fungicides containing Triademinol

686 Bayleton® BM

A wettable powder systemic fungicide containing 12.5% w/w triadimenol and 25% w/w carbendazim; used to control eyespot, mildew and early attacks of yellow and brown rust on winter wheat and winter barley, rhynchosporium on winter barley and eyespot and mildew on winter rye. *Bayer AG*. See triadimenol, carbendazim.

687 Baytan

Powder mixture of fuberidazole and triadimenol; seed dressing for cereals. *ICI Agrochemicals*. See fuberidazole, triadimenol.

688 Baytan® IM

Dry powder mixture of fuberidazole, imazalil and triadimenol; seed dressing for barley. *Bayer plc*. See fuberidazole, imazalil, triadimenol.

Fungicides containing Triadimefon

689 Bayleton® CF

Wettable powder fungicide with contact and systemic properties containing 6.25% w/w triadimefon and 65% w/w captafol; used to control powdery mildew, yellow and brown rust, leaf spot and glume blotch and to reduce the late-season ear disease complex on spring and winter wheat. *Bayer plc*. See triadimefon, captafol.

690 Dorin

triadimenol-tridemorph
Dorindan
Emulsifiable concentrate containing 125 g triadimenol and 375 g tridemorph per liter; used for control of mildew and rust in cereals. *Bayer plc*. See triadimenol, tridemorph.

691 Dorindan

Triadimenol + tridemorph.
Emulsifiable concentrate containing 125 g triadimenol and 375 g tridemorph per liter; used for control of mildew and rust in cereals. *Bayer plc*. See triadimenol, tridemorph.

Fungicides containing Tridemorph

692 Ashlade Blight Fungicide

Mixture of cymoxanil and mancozeb; used to control potato blight. *Ashlade Formulations Ltd*. See cymoxanil, mancozeb.

693 Rocket® Ultra

Tridemorph, fenpropiomorph; systemic fungicide for control of cereal diseases. *BASF AG*.

Fungicides containing Zinc

694 Amazin®

Maneb with zineb; fungicide. *Aceto*. See maneb, zineb.

695 Manex

Suspension concentrate containing maneb and zinc; a protectant fungicide against potato blight, mildew and rust. *Chiltern Farm Chemicals Ltd; Griffin; Quadrangle Agrochemicals; L W Vass (Agricultural) Ltd*. See maneb.

696 Mazin®

Wettable powder containing maneb 80% w/w and zinc oxide; protective fungicide against potato blight. *Universal Crop Protection Ltd*. See maneb, zinc oxide.

697 Patafol Plus

Manganese zinc ethylenebisdithiocarbamate (maneb, zineb) and ofurace (Patafol) mixture; for control of potato blight. (Sold in UK for Chevron Chemical Co.). *ICI Chem & Polymers Ltd*. See maneb, zineb, Patafol.

698 Tinman

Mixture of fentin hydroxide, maneb, and zineb; used for control of potato blight. *Chiltern Farm Chemicals Ltd*.

699 Trimanzone

Mixture of ferbam, maneb and zineb; a fungicide. *Pennwalt Chemicals Ltd*.

Miscellaneous Fungicides

700 Activex

Fungicides for garden use. *ICI Garden Products*.

701   Albricide
Fungicide. *Albright & Wilson Ltd., Phosphates & Specialty Business.*

702   Antiblu/Antiboror
Fungicide/insecticide; Used for freshly sawn timber. *Hickson & Welch Ltd.*

703   Armillatox®
Emulsion of polyhydric phenols in soap; home garden lawn treatment and fungicide; controls moss in lawns; reduces the severity of club root; hinders the spread of honey funfus. *Armillatox Ltd.*

704   Azosan
A fungicide. *May & Baker Ltd.*

705   Bardew
Fungicide. Schering Agrochemicals Ltd.

706   BAS 46402F
Systemic fungicide for use in winter wheat and barley. *BASF plc.*

707   Baymat®
Fungicide with good penetrant, protective, curative and eradicative activity for control of scab and blossom wilt on pome and stone fruit, rust, leafspot diseases and mildews on pome and stone fruit, bananas, vegetables, sugar beet, ornamentals. *Bayer AG.*

708   Binab T
Trichoderma viride.
Biological fungicide used to control silver leaf in fruit trees. *Henry Doubleday Research Association.*

709   Biobor
Diesel fuel fungicide. *U.S. Borax & Chem.*

710   Birgin®
Sprout suppressant for potatoes in storage. *Bayer AG.*

711   Bronocot
Cotton seed dressings. *ICI Chem & Polymers Ltd.*

712   Burcop
Soda/Bordeaux fungicide. *McKechnie Chemicals Ltd.*

713   Byacin
Iodophor in liquid solution; for controlling storage rots in potatoes. *Wheatley Chemical Co Ltd.*

714   Byatran
Iodophor and TBZ as dry granules; for controlling soil borne diseases in growing potato crops. *Wheatley Chemical Co Ltd.*

715   Calprona K
Mold inhibitor. *BP Chemicals Ltd.*

716   Cerere
Tricresylmercuroacetate.
A mixture of mono and diacetate derivatives of the three cresols, with about 75% of the mono-derivative, and containing about 57% mercury. It accelerates the germination of grain and affords protection against animal and vegetable parasites.

717   Clerit
Horticultural fungicide. *ICI Chem & Polymers Ltd.*

718   Clerite
Horticultural fungicide. Plant Protection.

719   Cobox® L
Ammonical copper polyacrylate.
Contact fungicide for control of fungus and bacterial diseases in coffee, cotton, fruits. *BASF AG.*

720   Comac Macuprax
Wettable powder containing 16.7% weight/weight copper sulfate and cufraneb (outside U.S.); a protectant fungicide against potato blight, canker in fruit trees and mildew in grapes. *McKechine Chemicals Ltd.*

721   Combined Seed Dressing
Fungicide/insecticide. *Murphy Chemical Co Ltd.*

722   Coopercote
Insecticidal varnish for paper, board etc. *The Wellcome Foundation Ltd.*

723   Copper Euparen®
3095
Copper + dichlorfluanid.
Combination product for prevention of fungal infections in fruit, grapes, vegetables and arable crops. *Bayer AG. See dichlorfluanid.*

724   Copper Green
Malachite.
A term applied to the mineral malachite. Used in pigments, pyrotechnics, insecticides, copper salts. An astringent in pomades, antidote for phosphorus poisoning, smut preventive, fungicide for seed treatment, feed additive. *See cupric carbonate.*

725   Copper-Lonacol
Copper + zineb.
Combination spray for the control of fungal diseases. *Bayer AG. See zineb.*

726 Coppesan
A copper fungicide. *The Boots Co plc.*

727 Cosmic® FL
Maneb + tridemorph + carbendazim.
Broad spectrum cereal fungicide. *BASF AG; BASF plc. See* tridemorph, carbendazim, maneb.

728 Dichlofuanide
N-Dimethylamino-N'-phenyl-N'-
(fluorodichlormethylthio) sulfamide.
A proprietary paint fungicide. *Bayer AG.*

729 Dikar
Fungicide and miticide. *Rohm & Haas.*

730 Dominate
A wettable powder containing a mixed culture of micro-organisms (*Anthrobacter sp., Aspergillus terreus, Bacillis subtilis, Bacillis thuringiensis, Bacteroides sp., Nocardia sp.,* and *Pseudomonas sp.*); used to suppress growth of pathognic soil fungi; applied to the soil; used for a wide variety of crops. *Westbridge Research Group.*

731 Dormakil
Fungicide (sold in UK for DowElanco). *ICI Chem & Polymers Ltd.*

732 Dufox
A potato fungicide. *Murphy Chemical Co Ltd.*

733 Farmaneb
Fungicide for prevention of potato blight. *ICI Chem & Polymers Ltd.*

734 FBC Protectant Fungicide
Fungicide. *Schering Agrochemicals Ltd.*

735 Fennite
A fungicide. *Fisons plc, Horticultural Div.*

736 Fernacol
A fungicide. *ICI Chem & Polymers Ltd.*

737 Fernasan
A nonmercurial seed dressing. *ICI Chem & Polymers Ltd.*

738 Ferox-Celotex
A proprietary trade name for Celotex which has been treated to resist attack by fungi and termites.

739 Florel Fruit Eliminator
Etephon.
Fungicide used to eliminate messy fruit set from ornamental olives, carobs, apples and crab apples; applied at bloom; also used to control mistletoe in conifers and deciduous trees. *Lawn & Garden Products Inc. See* etephon.

740 Focal
Flowable fungicide. *Schering Agrochemicals Ltd.*

741 Fomac
A fungicide for use with rubber. *ICI Chem & Polymers Ltd.*

742 Fungitrol Tinox
Bis(tri-n-butyl)oxide.
Wood preservative and antifouling coating compositions. *Hüls Am.*

743 Fungus Fighter
Fungicide. *May & Baker Ltd.*

744 Hysede
Lindane + thiabendazole + thiram.
Mixture of lindane, thiabendazole and thiram; seed dressing for brassica crops. *Agrichem (International) Ltd. See* lindane, thiabendazole, thiram.

745 Intercide
Biocide/fungicide for PVC. *AKZO Chemie UK Ltd.*

746 Isotox
Insecticide seed treater. *Monsanto (Solaris).*

747 Larvex
A solution of sodium fluosilicate; a proprietary clothes-moth remedy.

748 Lastil
Fungicide for bonded cork. *BDH Chemicals Ltd.*

749 Lonacol
Copper-free fungicide; used for tomatoes, beans, potatoes, maize, onions and fruits. *Bayer AG.*

750 Manox
Iron blue pigments; for printing inks, paints and fungicides. *Manox Ltd.*

751 Meld
Broad spectrum systemic fungicide for use in cereals. *BASF plc.*

752 Merquinox
Fungicide for industrial applications. *Octel Chemicals Ltd.*

753 Mersil
Turf fungicide. *May & Baker Ltd.*

754   Mesgamma
Combined insecticide fungicide and seed dressing. *Plant Protection.*

755   Mi-Col
A mildew fungicide. *Plant Protection.*

756   Microsan
A copper fungicide. *Mechema Chemicals Ltd.*

757   Mist-O-Matic
A liquid fungicide seed dressing. *DowElanco Ltd.*

758   Mist-o-matic Ferrax
Fungicide seed treatment. *Murphy Chemical Co Ltd.*

759   Mortegg Emulsion
Emulsifiable concentrate containing 600 g/l tar oils; fungicidal winter wash for fruit. *DowElanco Ltd.*

760   Moss Gunl
Ready-to-use mosskiller in spray form. *ICI Garden Products.*

761   Mos-Tox
Moss eradicant. *May & Baker Ltd.*

762   Neantina
Seed dressing for control of fungal diseases on cereals, rice, cotton and vegetables. *Bayer AG.*

763   Occidine
Contains copper sulfate, iron sulfate, sulfur, naphthalene, and calcium carbonate; used as a fungicide.

764   Orisan
Seed dressing for control of fungal diseases on cereals, rice, cotton and vegetables. *Bayer AG.*

765   Pallitop® S
Nitrothal + isopropyl-sulfur.
Contact fungicide against powdery mildew in apples. *BASF AG. Seepyl,* sulfur.

766   Panazyme
Fungal protease. *Rhône-Poulenc UK.*

767   Paramos
Benzalkonium chloride.
Algicide and moss killer for paths and flower pots. *Chemsearch (UK) Ltd. See* benzalkonium chloride.

768   Pentol
Timber fungicides. *Plant Protection.*

769   Persulon
Fungicide; used for control of powdery mildew on cereals, fruit, vegetables and ornamentals. *Bayer AG.*

770   Preventol® A2
An organic inhibitor; for use in the formulation of fungicidal interior paints (emulsion and solvent based paints) excluding air-drying paints. *Bayer AG.*

771   Protars
A dry arsenical fungicide prepared from talc, lime, and arsenic oxide.

772   Red Storax
Solid storax.
An artificial product obtained by mixing poor storax with sawdust, and pressing the mixture; used for fumigating candles and powders. *See* storax.

773   RH Maneb 80
Protectant fungicide. *Rohm & Haas UK. See* maneb.

774   Rizolex
Tolclofos + methyl.
Fungicide. *Schering Agrochemicals Ltd.*

775   Roseclear
bupirimate + pirimicarb + triforine.
Contains bupirimate, pirimicarb and triforine; combined insecticide and fungicide for garden use. *ICI Garden Products. See* bupirimate, pirimicarb, triforine.

776   Rovral
Fungicide. *Rhône-Poulenc Rorer Ltd. See* iprodione.

777   Rovral WP
Horticultural fungicide. *Embetec Crop Protection Ltd. See* iprodione.

778   Screen
Seed protectant. *Monsanto Co.*

779   Sofanate
A fungicide for fruit storage. *Plant Protection.*

780   Stanza
Fungicide. *Schering Agrochemicals Ltd.*

781   Super AD-IT
Di(phenylmercuric) dodecenyl succinate.
Preservative and fungicide for aqueous coating compositions. *Hûls Am.*

782   Systemic Fungicide
Garden fungicide. *Murphy Chemical Co Ltd.*

783   Tecane
Selective herbicide. *Schering Agrochemicals Ltd.*

784   Terminate
Bacillus thuringiensis wettable powder; applied by spray to control larvae of *Lepidopteran* insects. *Westbridge Research Group.*

785   Tillantina
Seed dressing for control of fungal diseases on cereals, rice, cotton, and vegetables. *Bayer AG.*

786   Trithac
Fungicide. *Murphy Chemical Co Ltd.*

787   Tubazole M
TBZ and iodophor as a sprayable solution; for controlling various diseases in stored potatoes. *Wheatley Chemical Co Ltd. See* thiabendazole.

788   Tumbleblite
Systemic fungicide. *Murphy Chemical Co Ltd.*

789   Zincofol
A fungicide. *Monsanto (Solaris).*

790   Zintox
Zinc arsenate.
A proprietary trade name for an agricultural spray containing basic zinc arsenate.

Herbicides

791   Acifluorfen
50594-66-6   111   256-634-5
$C_{14}H_7ClF_3NO_5$
5-[2-Chloro-4-(trifluoromethyl)phenoxy]-2-nitrobenzoic acid.
(sodium salt) scifluorfen; RH-6201; Blazer. Herbicide. mp = 151-157°; (sodium salt): mp = 124-125°; soluble in H_2O (>25 g/100 ml); LD_{50} (rat orl) = 1300 mg/kg.

792   Acifluorfen Sodium Salt
62476-59-9   263-560-7
$C_{14}H_6ClF_3NNaO_5$
5-(2-Chloro-4-(trifluoro-methyl)phenoxy)-2-nitro-benzoic acid sodium salt.
sodium acifluorfen; Tackle; Blazer®. For post-emergence control of broad-leaved weeds and suppression of some annual grasses in soybeans. *BASF AG. See* acifluorfen.

793   Agriphlan 24
1582-09-8   9815   216-428-8
$C_{13}H_{16}F_3N_3O_4$
2,6-Dinitro-N,N-dipropyl-4-(trifluoromethyl)benzeneamine.
Trifluralin; trifluraline; Brassix, Digermin; Elancolan; Heritage; Ipersan; L-36352; Olitref; Proflan; Prolan; Sinflouran; Tarene; Tri-4; Triflurex; Trigard; Trimaran; Tristar; Zeltoxone; Ornamental Weeder; Treflan; Trigard; Trilin® 10G; Trimaran; Tripart® Trifluralin 48 EC; Tristar. A herbicide against cottonweeds, bean, tomato, pear, garlic and sunflower weeds; very efficient against amaranth, bristle-grass, knapweed etc. A preemergence herbicide; granular material for use on ornamentals, shrubs, trees, roses, and flower beds; control grasses and many broadleaf weeds. Also used for the control of certain germinating broad-leaved weeds and annual grasses in beans brassicae, cabbage, and a wide variety of other crops. *Chemical Combine; Lawn & Garden Products Inc.; DowElanco Ltd.; Farmers Crop Chemicals Ltd.; Griffin; Ashlade Formulations Ltd.; Tripart Fram Chemicals Ltd.; Pan Britannica Industries Ltd. See* trifluralin.

794   Alachlor
15972-60-8   203   240-110-8
$C_4H_{20}ClNO_2$
2-Chloro-N-(2,6-diethylphenyl)-N-(methoxymethyl)acetamide.
Alanex; Alagan; Alazine; 2-chloro-2',6'-diethyl-N-(methoxymethyl)acetanilide; Lasso; Metachlor; CP-50144; alachlor-atrazine. Active ingredient: alachlor; pre-emergence and pre-plant incorporated herbicide for control of most annual grasses and certain broadleaf weeds. Herbicide mp = 40-41°; $bp_{0.3}$ = 135°; soluble in H_2O, organic solvents; LD_{50} (rat, orl) = 1200 mg/kg. *Agan Chemical Manufacturers Ltd.*

795   Alloxydim Sodium
55635-13-7   259-733-1
$C_{17}H_{24}NNaO_5$
Sodium 2,2-dimethyl-4,6-dioxo-5-(1-((2-propenyloxy)imino)butyl)cyclohexanecarboxylate ion(1-).
Clout; Tritex; Kusagard; Fervin; ADS; NP-48; NP-48NA; Sodium alloxydim. Herbicide for control of annual weeds. *Embetec Crop Protection Ltd.*

796   Amcide
7773-06-0   589   231-871-7
$H_6N_2O_3S$
Ammonium sulfamate.
Fergon; Root out. An inorganic herbicide to control weeds and grasses in vegetables and ornamentals prior to planting and as a tree, weed and brushwood killer. *Battle, Hayward & Bower Ltd.; Dax Products Ltd. See* ammonium sulfamate.

797   Ametryn

834-12-8   411   212-634-7

$C_9H_{17}N_5S$

N-Ethyl-N'-(1-methylethyl)-6-(methylthio)-1,3,5-triazine-2,4-diamine.

Ametrex; ametryne; Amephyt; Ametrex; Doruplant; Evik; G 34162; Gesapax; Mebatryne; Evik 80W; Cemerin; 2-Ethylamino-4-isopropylamino-6-methylthio-s-triazine; Amigan; Ametrex. Active ingredient: ametryn; selective pre- and post-emergence herbicide, also used as an aquatic herbicide and vine desiccant. Selective systemic herbicide used for control of most annual grasses and broad-leaved weeds in pineapples, sugar cane, bananas; citrus fruit, maize, cassava; coffee, tea, cocoa; oil palmsand on non-crop land. An unrestricted, general use pesticide. mp = 84-85°; d_{20} = 1.19; soluble in H_2O (185 mg/l), more soluble in organic solvents; LD_{50} (rat orl) = 1110 mg/kg. *Agan Chemical Manufacturers Ltd.*

798   Aminotriazole Bayer

61-82-5   513   200-521-5

$C_2H_4N_4$

1,2,4-Triazol-3-ylamine.

Amitrole; Azolan; Boroflow S/ATA; MSS Aminotriazole; Primatol SE 500FW; 3-Amino-s-Triazole;ATA; Amizol; Amerol; Azolan; Herbizole; Amino Triazole Weedkiller 90; AT-90; amitol; cytrole; domatol; emisol; triazolamine; Vorox; ramizol; 3,A-T; zaplant; weedar ads; weedazin; weedoclor; aminotriazol-spritzpulver; azaplant kombi; campaprim a 1544; elmasil; herbidal total; kleer-lot; orga-414; radoxone tl; weedex granulat; X-ALL; aminotriazole (plant regulator); amitril t.l.; amitrol 90; amizol d; amizol dp nau; amizol f; domatol 88; emisol 50; emisol f; fenavar; vorox aa; vorox as; weedar at; weedazin arginit; weedazol gp2; weedazol super; weedazol t; Weedazole; Diurol; Triazol-3-amine. Fast-acting herbicide, for control of hard-to-kill grass and broad-leaved weeds; mainly used in mixtures with other compounds. Weedkiller with good translocation characteristics for the control of perennial and annual weeds. *Bayer AG; Agan Chemical Manufacturers Ltd.; ABM Chemicals Ltd.; Mirfield Sales Services Ltd.; Ciba-Geigy Agrochemicals.* See amitrole.

799   Ammonium Thiocyanate

1762-95-4   597   217-175-6

CH_4N_2S

Ammonium rhodanide.

Thiocyanic acid, ammonium salt; Ammonium sulfocyanate; Ammonium Sulfocyanide; Ammonium Rhodantate; Ammonium Rhodonide; Ammonium Rhodanide; ATC; Trans-aid. Analytical chemistry; thiourea; fertilizers; photography; in liquid rocket propellants; fabric dyeing; zinc coating; weed killer, defoliant;

adhesives; curing resins; pickling iron and steel; electroplating; polymerization catalyst; metals separation. mp = 149°; d = 1.3050; soluble in H_2O (163 g/100 ml); LD_{50} (rat orl) = 750 mg/kg. *Carbo-Tech GmbH; Degussa; Witco/Argus.*

800   Aresin

1746-81-2   217-129-5

$C_9H_{11}ClN_2O_2$

N'-(4-Chlorophenyl)-N-methoxy-N-methylurea.

Monolinuron; Afesin; Arresin; Monorotox. Selective systemic herbicide used to control broad-leaved weeds and some annual grasses in vegetable crops such as potatoes and leeks. Used as an emulsifiable concentrate containing 200 g/l monolinuron, for control of annual dicotyledons in potatoes, french beans and leeks. mp = 80-83°; soluble in H_2O (735 mg/l), organic solvents; LD_{50} (rat orl) = 1660 mg/kg. *Hoechst UK.*

801   Arsenic Trioxide

1327-53-3   844   215-481-4

As_4O_6

Arsenic (III) oxide.

Arsenic oxide;Arsenous trioxide; arsenous acid; arsenous oxide; arsenic sesquioxide; White Arsenic; Diarsenic Trioxide; Crude Arsenic; Arsenic (white); Arsenious oxide; Arsenic (III) trioxide; Arsenous anhydride; arsenite; arsenolite; arsenous acid anhydride; arsenous oxide anhydride; arsodent; claudelite; claudetite; Arsenic oxide (3); Arsenic oxide (As_2O_3); Arsenic sesquioxide (As_2O_3); Arsenicum album; Diarsonic trioxide; Diarsenic oxide. Pigments, ceramic enamels, aniline colors, decolorizing agent in glass, insecticide, rodenticide, herbicide, sheep and cattle dip, hide preservative, wood preservative, preparation of other arsenic compounds. mp = 315°; bp = 465°; soluble in H_2O, dil HCl, alkali hydroxide or carbonate solns; insoluble in EtOH, $CHCl_3$, Et_2O; LD_{50} (rat orl) = 1.46 mg/kg. *Atomergic Chemetals; Noah Chem.; Outokumpu Oy; Transene.*

802   Asulam

3337-71-1   222-077-1

$C_8H_{10}N_2O_4S$

Methyl ((4-aminophenyl)sulfonyl)carbamate.

Norunil; Asulox; Asulox 3.34L; Asilan; Asulox 40; Asulox F; Jonnix; MB 9057; Methyl sulfanilylcarbamate. A soluble concentrate containing 400 g asulam per liter; a selective weedkiller and herbicide for control of docks and bracken. *Embetec Crop Protection Ltd; Rhône-Poulenc Environmental Prods. Ltd.; May & Baker Ltd.* See asulox.

803 Atrazine

1912-24-9 902 217-617-8

$C_8H_{14}ClN_5$

6-Chloro-N-ethyl-N'-(1-methylethyl)-1,3,5-triazine-2,4-diamine.

Ashlade 4% At Gran; Ashlade Atrazine 50 FL; Atraflow; Atranex; Borocil A; Boroflow A; Boroflow A/ATA; Chlorea; Gesaprim 500FW; Herbazin Total; Mascot Gauntlet; Primatol AA; Atraflow; Atranex; Ashlade 4% At Gran; Atraflow Plus; Atrex; Atratol; Primatol A; A 361; Aatrex; Aktinit A; G 30027; Gesaprim; Hungazin; Atranex; Fogard; Griffex; Mebazine; Vectal; Atrazines; Extrazine II; Laddock; Aatrex 4l; Aatrex 80W; Atrazine 4l; Atrazine 80W; Griffex 4l; Ortho St. Augustine Weed and Feed; Scotts Bonus Type S; Crisazina; Vectral SC; Attrex; Crisamina; Vectal SC; ATZ; Triazine A 1294; Zeazin; Argezin; Aktikon; Aktikon PK; Aktinit PK; Wonuk; Oleogesaprim; Chromozin; Pitezin; Actinite PK. Active ingredient: atrazine. Selective systemic herbicide, absorbed through roots and foliage; inhibits photosynthesis. Used for pre- and post-emergence control of annual grasses and broad-leaved weeds in a variety of crops. mp = 176°; d = 1.38; soluble in H_2O (28 mg/l), more soluble in organic solvents; LD_{50} (rat orl) = 1100, 3080 mg/kg. *Ashlade Formulations Ltd.; Rhône-Poulenc Environmental Prods. Ltd.; Agan Chemical Manufacturers Ltd.; ABM Chemicals Ltd.; Chipman Ltd.; Ciba-Geigy Agrochemicals; Fisons plc, Horticultural Div.; Rigby Taylor Ltd.; Ciba plc.*

804 Aziprotryn

4658-28-0 225-101-9

$C_7H_{11}N_7S$

4-Azido-N-(1-methylethyl)-6-methylthio-1,3,5-triazin-2-amine.

C 7019; Mesoranil; Brasoran; Brasoran 50 WP; Mesanoril. A selective herbicide used to control a wide range of annual broad-leaved weeds and some grasses in brassicas. *Ciba-Geigy Agrochemicals.*

805 Benazolin

3813-05-6 223-297-0

$C_9H_6ClNO_3S$

4-Chloro-2-oxo-3(2H)-benzothiazoleacetic acid.

Eunasin; Cresopur; Benzar; BTS-7693; Cornox; BEN-30; Ben-cornox; Benopan; Bensecal; Cornox CWK; EX 10781; Galipan; Grassland weedkiller; Herbazolin; Keropur; Legumex extra; Ley-cornox; Leymin; Metizolin; Tri-cornox special. Selective, systemic growth regulator herbicide Used for control of broad leaved weeds such as black bindweed, chickweed, cleavers and charlock. mp = 193°; soluble in H_2O (600 mg/l), more soluble in organic solvents; LD_{50} (rat orl) > 4800 mg/kg. *Schering Agrochemicals Ltd.*

806 Benfluralin

1861-40-1 1067 217-465-2

$C_{13}H_{16}F_3N_3O_4$

N-Butyl-N-ethyl-2,6-dinitro-4-trifluoromethylaniline.

Benefex; Benefin; EL-110; Balan; Balfin; Quilan. Active ingredient: benfluralin. A pre-emergence herbicide with a wide range of weed control both of annual grass weeds and broad-leaved weeds. Yellow-orange crystalline solid; mp = 65-66.5°; soluble in most organic solvents. *Agan Chemical Manufacturers Ltd. See balan.*

807 Bentazon

25057-89-0 1080 246-585-8

$C_{10}H_{12}N_2O_3S$

3-(1-Methylethyl)-1H-2,1,3-benzothia-diazin-4(3H)-one 2,2-dioxide.

Basagran®; bentazone; BAS 351H; Adagio; Galaxy; Storm; Basagran 4E; Thiadiazinol; Bendioxide. A contact herbicide. mp = 137-139°; soluble in H_2O; LD_{50} (rat orl) = 1100 mg/kg. *BASF plc.*

808 Bidisin

101-10-0 202-915-2

$C_9H_9ClO_3$

DL-2-(3-Chlorophenoxy)-propionic acid.

cloprop; Amchem 3-CP; Bidisin forte; Metachlorphenprop. Used for control of wild oats in cereals, maize and sugar beet. *Rhône-Poulenc. See cloprop.*

809 Bifenox

42576-02-3 1256 255-894-7

$C_{14}H_9Cl_2NO_5$

5-(2,4-Dichlorophenoxy)-2-nitrobenzoic acid methyl ester.

MC-4379; Modown; Modown 4 Flowable. Selective herbicide used for control of annual broad-leaved weeds and some grasses in cereals, maize, sorghum, soybeans, rice and some other crops. mp = 84-86°; poorly soluble in H_2O (0.43 mg/l), more soluble in organic solvents; LD_{50} (rat orl) > 6400 mg/kg.

810 Bifenthrin

82657-04-3 1257

$C_{23}H_{22}ClF_3O=72$

1α,3α(Z)-(±)-3-(2-Chloro-3,3,3-trifluoro-1-propenyl)-2,2-dimethylcyclopropanecarboxylic acid.

Talstar; biphenate; FMC 54800; Brigade; Capture; Capture 2; Biflex. An emulsifiable concentrate containing 100 g bifenthrin per liter; a residual herbicide for the control of weeds in winter cereals, oilseed rape and peas. *DowElanco Ltd.*

811   Bromacil

314-40-9   1402   206-245-1

$C_9H_{13}BrN_2O_2$

5-Bromo-6-methyl-3-(1-methylpropyl)-
2,4(1H,3H)pyrimidinedione.

5-bromo-3-sec-butyl-6-methyluracil;   5-bromo-6-
methyl-3-(1-methylpropyl)uracil; Borea; Bromax;
Croptex Onyx; Du Pont Herbicide 976; Hyvar X;
Nalkil; Rokar X; Rout; Staa-Free; Urox B; Uragan;
Du Pont Herbicide 976; Hyvar X; Nalkil; Rokar
Xrout; Staa-Free; Uragan; Urox B; Hyvar X;
Hyvar® X. A versatile herbicide for control of
established annual and perennial broadleaf weeds
and grasses and brush. Photosynthesis inhibitor,
used as a herbicide for total weed and brush
control on non-crop land and selective control of
annual and perennial weeds and grasses in citrus
and pineapple plantations. A wettable powder
containing 80% w/w bromacil; used for control of
weeds in cane fruit and noncrop areas. mp = 158-
169°; d^{25} = 1.55; soluble in H_2O (815 mg/l), less
soluble in organic solvents; LD_{50} (rat orl)= 5200
mg/kg. *Agan Chemical Manufacturers Ltd.;
Selectokil Ltd.; DuPont UK.*

812   Bromoxynil

1689-84-5   1465   216-882-7

$C_7H_3Br_2NO$

3,5-Dibromo-4-hydroxybenzonitrile.

Brominal; Bromotril; Buctril; Certrol B; Litarol;
M&B 10064; Merit; Pardner; Sabre; Torch.
Selective weedkiller. A selective contact herbicide
used for post-emergence control of some annual
broad-leaved weeds in cereal crops. mp = 84°;
soluble in H_2O (50 mg/l), organic solvents; LD_{50}
(rat orl) = 365 mg/kg. *May & Baker Ltd.*

813   Butachlor

23184-66-9   1533   245-477-8

$C_{17}H_{26}ClNO_2$

N-(Butoxymethyl)-2-chloro-2',6'-
diethylacetanilide.

2-chloro-2,6-diethyl-N-(butoxymethyl)acetanilide;
Butanex; CP-53619; Machete. Active ingredient;
selective pre-emergence and early post-emergence
weed control in transplanted, direct seeded and
upland rice. $bp_{0.5}$ = 196°; d_4^{30}= 1.0695; soluble in
H_2O (20 mg/l), organic solvents; LD_{50} (rat orl) =
1740 mg/kg. *Agan Chemical Manufacturers Ltd.;
Monsanto Co.*

814   Butisan® S

67129-08-2   266-583-0

$C_{14}H_{16}ClN_3O$

2-Chloro-N-(2,6-dimethylphenyl)-N-(1H-pyrazol-
1-ylmethyl)acetamide.

metazachlor; Butisan S; metazachlore; BAS
47900H; Track; Pree®. Selective herbicide,
inhibits germination. Used for control of annual
grasses and broad-leaved weeds in fruit and
vegetable crops. Systemic herbicide against
annual grasses, broadleaf weeds, brassicas, maize
and ornamental crops. mp = 85°; soluble in H_2O
(17 mg/l), freely soluble in organic solvents; LD_{50}
(rat orl) = 2150 mg/kg. *BASF AG; Bayer plc;
Kommer-Brookwick Ltd. See metazachlor.*

815   Cetrimide

57-09-0

$C_{19}H_{42}BrN$

Hexadecyltrimethylammonium bromide.

Cetab; Centimide; Cetyltrimethylammonium
Bromide; HTAB; CTAB; Cetrimonium bromide;
CTABr; Cetrimide. Not a herbicide. Surfactant
used to apply herbicides. mp = 218°. *Alfa Aesar;
Fisher; Lancaster Synthesis Inc.; TCI; Pfaltz &
Bauer; Acros Organics - USA; ICN Biomedical
Research Products.*

816   Chloridazon

1698-60-8   216-920-2

$C_{10}H_8ClN_3O$

5-Amino-4-chloro-2-phenyl-3(2H)-pyridazinone.

pyrazon; PAC; PCA; Weedmaster; Alicep®; Atlas
Silver; Better Flowable; Chiltern Pyrazol;
Gladiator; Paramin DF; Pyramin®; Starter
Flowable; Tripart® Gladiator; Trojan SC; Better;
Brek; Curbetan; Gladiator; H 119; Hyzon;
Pyramin; Pyrazol; Silver; Starter; Trojan. Selective
systemic herbicide, rapidly absorbed by roots,
used for control of annual broad-leaved weeds in
sugar beets, fodder beet and beetroot. Used for
pre- and post-emergence weed control in onions,
leeks, chives, and flower bulbs. mp = 205-206°
(dec); soluble in H_2O (400 mg/l), more soluble in
organic solvents; LD_{50} (rat orl) = 3830 mg/kg.
*Sipcam; Agrimont; Wacker; Tripart Farm
Chemicals Ltd.; BASF; Atlas Interlates, Truchem;
Portman Agrochemicals Ltd.; BASF AG; BASF plc;
Atlas Interlates Ltd.; Chiltern Farm Chemicals Ltd.;
Truchem Ltd.; Schering Agrochemicals Ltd.;
Rhône-Poulenc Environmental Prods. Ltd.;
Agrichem (International) Ltd.; Farmers Crop
Chemicals Ltd.; MTM AgroChemicals Ltd.*

817   Chlorotoluron

15545-48-9   239-592-2

$C_{10}H_{13}ClN_2O$

N-(3-Chloro-4-methylphenyl)-N',N'-dimethylurea.

Ludorum; Dicurane 500 FW; Toro; Tripart®
Ludorum 700; Dicurane 500 FW; Toro; Tripart®
Ludorum 700; Dicuran; Tolurex; Chlortokem.
Suspension concentrate containing 500 g
chlorotoluron per liter; a contact urea herbicide
for cereal crops. Chlortoluron 700 g/liter (58.5%
w/w); used for the control of black grass, wild
oats, and other annual grasses and a range of
broad-leaved weeds in a range of named wheats,
winter barleys, durum wheats and triticale. *Tripart
Farm Chemicals Ltd; Sipcam UK Ltd.; Tripart Farm*

Chemicals Ltd.; Ciba-Geigy Agrochemicals; Sipcam UK Ltd.

818 Chloroxuron

1982-47-4 217-843-7

$C_{15}H_{15}ClN_2O_2$

N-[4-(4-Chlorophenoxy)phenyl]-n,n-dimethylurea. chloroxyfenidim; chlorphencarb; C 1983; Gesamoos; Tenoran. Selective herbicide, inhibits photosynthesis. Used for pre- and post-emergence control of annual broad-leaved weeds and some grasses in peas, beans, carrots, celery, onions, leeks, garlic, chives, fennel, parsley, dill, tomatoes, cucurbits, soya beans Urea herbicide for on strawberries and ornamentals. mp = 151-152°; soluble in H_2O (3.7 mg/l), more soluble in organic solvents; LD_{50} (rat orl) = 3700 mg/kg. *Ciba-Geigy Agrochemicals.*

819 Chlorpropham

101-21-3 2240 202-925-7

$C_{10}H_{12}ClNO_2$

(3-Chlorophenyl) carbamic acid 1-methylethyl ester.

Carbanilic acid, m-chloro- isopropyl ester; (3-chlorophenyl)carbamic acid, 1-methylethyl ester; Beet-Kleen; Chloro-IPC; Chloro-IFK; Chloro IPC; Chlorpropham; Chlorpropam; Chlorpropam; Chlorpropham; ChlorIPC; CI-IPC; Elbanil; ENT 18,060; Fasco WY-HOE; Furloe; Atlas CIPC 40; Furloe 3 EC; isopropyl chlorocarbanilate; IPC, CIPC; Croptex Chrome; Campbell's CIPC 40%; Warefog; Residuren; Spud-Nic®; MSS CIPC. Emulsifiable concentrate containing 400 g chlorpropham per liter; a carbamate herbicide. Selective systemic herbicide and growth regulator. Used for pre-emergence control of annual grasses and broad-leaved weeds. A carbamate herbicide and sprout depressant in stored potatoes. mp = 40.7-41.1°; bp$_2$ = 149°; d$_{30}$= 1.180; soluble in H_2O (89 mg/l), mores soluble in organic solvents; LD_{50} (rat orl) = 1200 mg/kg. *Hortichem Ltd.; Pfizer International; Atlas Interlates Ltd.; Wheatley Chemical Co Ltd; Dean Agrochemicals Ltd.; MTM AgroChemicals Ltd.; ICI Chem & Polymers Ltd.; Aceto; Mirfield Sales Services Ltd.*

820 Clarosan 1FG

886-50-0 212-950-5

$C_{10}H_{19}N_5S$

N-(1,1-Dimethylethyl)-N'-ethyl-6-(methylthio)-1,3,5-triazine-2,4-diamine.

A 1866; Clarosan; Igran; Igran 50; Igran 500; Prebane; Prebane 500; Shortstop; Shortstop E; Terbutrex; Terbutryn; Terbutryne; Athado; GS 14260; Plantonit. Selective herbicide absorbed by roots and foliage and used for control of most grasses in winter cerals, vegetables and citrus fruit. Used for pre-emergence and post-emergence weed control in cererals and aquatic weed control. mp = 104-105°; bp$_{0.06}$ = 154-160°; d$_{20}$ = 1.115; soluble in H_2O (25 mg/l 20°), more soluble in organic solvents; LD_{50} (rat orl) = 2045 mg/kg. *Ciba-Geigy AgroChemicals; Probelte; Ciba-Geigy; Chemolimpex; Makhteshim-Agan; Agan Chemical Manufacturers Ltd. See terbutryn.*

821 Clopyralid

1702-17-6 2462 216-935-4

$C_6H_3Cl_2NO_2$

3,6-Dichloropyridine-2-carboxylic acid.

Benazalox; Cirtoxin; Cyronal; Dowco 290; Lontrel; Matrigon; Reclaim; Shield Stinger; CDA Dicotox Extra; Agrichem; Agricorn D; BH 2,4-D Ester 40; Campbell's Destox; Dow Shield; Format; Shield; Silvapron. Post-emergence control of broad-leaf weeds of *Polygonaceae, Compositae, Leguminosae* and *Umbelliferae*. Good control of creeping thistle, sow thistle, coltsfoot, mayweeds and *Polygonum* spp. Translocated herbicide for cereals and established grassland. A foliar herbicide for use on brassicas and field vegetables. mp = 151-152°; soluble in H_2O (9 g/l), soluble in organic solvents; LD_{50} (rat orl) >4300 mg/kg. *DowElanco Ltd.; ICI Chem & Polymers Ltd.; Murphy Chemical Co Ltd,; BP Oil Ltd.*

822 Cyanazine

21725-46-2 2755 244-544-9

$C_9H_{13}ClN_6$

2-[[4-Chloro-6-(ethylamino)-1,3,5-triazin-2-yl]amino]-2-methylpropanenitrile.

Propionitrile, 2-[[4-chloro-6-(ethylamino)-1,3,5-triazin-2-yl]amino]-2-methyl; s-triazine, 2-chloro-4-(ethylamino)-6-(1-cyano-1-methyl)(ethylamino)-; Bladex; 80WP; Cyanazine SD 15418; DW 3418; Fortol; Fortrol; Payze; SD 15418; WL 19805; Match. Suspension concentrate containing 500 g cyanazine per liter; a triazine herbicide. Selective systemic herbicide, absorbed by roots and foliage. Used for control of annual grass and broad-leaved weeds. mp = 167.5-169°; soluble in H_2O (171 mg/l), more soluble in organic solvents; LD_{50} (rat orl) = 149 mg/kg. *DuPont; Shell UK.*

823 Cycloxydim

101205-02-1

$C_{17}H_{27}NO_3S$

2-(1-(Ethoxyimino)butyl)-3-hydroxy-5-(tetrahydro-2H-thiopyran-3-yl)-2-cyclohexene-1-one.

Laser; Stratos®; Focus®; BAS-517H. Selective herbicide, absorbed primarily by leaves. Inhibits mitosis. Used for post-emergence control of annual and perennial grasses in broad-leaved crops, such as beans and potatoes. Post-emergence graminicide against annual and perennial grasses; selective in broadleaf crops, e.g., sugar beet, cotton, soybean, vegetables, onions mp = 36°; soluble in H_2O (88 mg/l), freely soluble in organic solvents; LD_{50} (rat orl)= 3940 mg/kg. *BASF AG.*

824   2,4-D
94-75-7   2865   202-361-1
$C_8H_6Cl_2O_3$
2,4-Dichlorophenoxyacetic acid.
Weedone-2,4-DP; 2,4-D Amine No. 4; 2,4-D LV6; Amine 4 2,4-D Weed Killer; Asgrow 2,4-D amine 4; Asgrow Aqua KD; DMA 4; Formula 40; Formula 40 4L; Helena 2,4-D; Low Vol 4 Ester Weed Killer; MCP amine; Standard 2,4-D Amine; Weedar 64; Weedar 64A; Agrotect; Amoxone; BH 2,4-D; Chloroxone; Crop Rider; Debroussaillant 600; Dormone; Emulsamine BK; Envert DT; Fernimine; Lawn-Keep; Miracle; Weed Tox; Weedtrol; amidox; b-selektonon; chipco turf herbicide ''d''; crotilin; decamine; dicopur; dicotox; ipaner; monosan; netagrone; pennamine; rhodia; U-5043; verton. Selective weed killer containing 465 g/l 2,4-D; For control of broadleaf weeds on amenity areas, golf courses, playing fields, etc. A translocatable herbicide for cereals and established grassland. *Burts and Harvey; Rhône-Poulenc Environmental Prods. Ltd.; Makhteshim Chemical Works Ltd.; Atlas Interlates Ltd.; MTM AgroChemicals Ltd.; Farm Protection Ltd.; ICI Chem & Polymers Ltd.; Synchemicals Ltd.; Mirfield Sales Services Ltd.; Shell UK.*

825   2,4-DB
94-82-6   2893   202-366-9
$C_{10}H_{10}Cl_2O_3$
4-(2,4-Dichlorophenoxy)butanoic acid.
2,4-DB; Butoxone; Campbell's DB Straight; Embutox; Embutone; Venceweed; Legumex; DB; Butirex; Butormone; Butoxone; Butyrac; M&B 2878. Selective systemic hormone type herbicide. Used for post-emergence control of many annual and perennial broadleaf weeds in lucerne, clovers, cereals, grassland, forest legumes, soybeans and ground nuts. Soluble concentrate containing 300 g 2,4-DB per liter; used to control weeds in lucerne. mp = 117-119°; soluble in H_2O (46 mg/l), more soluble in organic solvents; LD_{50} (rat orl) = 370-700 mg/kg. *ICI Chem & Polymers Ltd.; MTM AgroChemicals Ltd.; May & Baker Ltd.*

826   Dacthal
1861-32-1   2896   217-464-7
$C_{10}H_6Cl_4O_4$
2,3,5,6-Tetrachloro-1,4-benzenedicarboxylic acid dimethyl ester.
DCPA;   2,3,5,6-tetrachloro-1,4-benzenedicarboxilic acid dimethyl ester; 2,3,5,6-tetrachloroterephthalic acid dimethyl ester; dimethyl 2,3,5,6-tetrachloroterephthalate; chlorthal-methyl; Rid; Dimethyl tetrachloroterephthalate; Vegetable Turf and Ornamental Weeder; Dacthal; DCPA; Chlorothal; Chlorthal-dimethyl; Chlorthal-methyl;

Chlorthal-Dimethyl; Dacthalor; Dimethyl ester of tetrachloroterephthalic acid; 2,3,5,6-tetrachloro-1,4-benzenedicarboxylate; DAC 4; DAC-893; DCP; DCPA; Fatal; Terephthalic acid, 2,3,5,6-tetrachloro-,dimethyl ester; Rid; Terechloro-terephthalic acid dimethyl ester. Wettable powder containing 75% w/w dimethyl tetrachloro-terephthalate (DCPA); an herbicide for use on ornamentals, turf, fruit and vegetables. Pre-emergence sprayable herbicide that can be used in vegetable gardens on ornamentals and in turf areas; controls spurge. mp = 155-156°. *Lawn & Garden Products Inc.; Fermenta ASC Europe Ltd.*

827   Dalapon
75-99-0   2869   200-923-0
$C_3H_4Cl_2O_2$
Sodium 2,2-dichloropropionate.
Couch and Grass Killer; Synchemicals Dalapon; Basfapon/N; BH dalapon; Basinex; Crisapon; Davpon; Gramevin; Kenapon; Uropon; Dalapon; Dowpon Proprop; Revenge; Unipon; Alatex; S95; Dalapon 85; Basfapon. Soluble powder containing dalapon; used for control of grasses in crop and noncrop areas. Selective systemic herbicide absorbed by roots and leaves. Used for control of annual and perennial grasses on non-crop land and also orchards and vineyards. bp_{18} = 98-100°; d = 1.4014; n_D^{20} = 1.4544; LD_{50} (rat orl) = 7126 mg/kg. *Rhône-Poulenc Environmental Prods. Ltd.; Vitax Ltd.; Synchemicals Ltd.*

828   Dekryll
534-52-1   3331   208-601-1
$C_7H_6N_2O_5$
3,5-Dinitro-2-hydroxytoluene.
4,6-dinitrocresol; Dinitro-o-cresol; DNOC; DNC; Nitrador; Dinitrocresol; Antinonnin; Detal; Dinitrol; Elgetol; K III; K IV; Ditrosol; Prokarbol; Effusan; Lipan; Selinon; Dekrysil; Antinonin; dinitrosol; Elgetox; 4,6-DNOC; Elgetol 30; Sinox. Proprietary preparation of 4,6-dinitrocresol, selective herbicide, insecticide. *See 4,6-dinitrocresol.*

829   Desmetryn
1014-69-3   213-800-1
$C_8H_{15}N_5S$
N-Methyl-N'-(1-methylethyl)-6-(methylthio)-1,3,5-triazine-2,4-diamine.
Semeron;   Isopropylamino-4-(methylamino)-6-(methylthio)-s-triazine; Methylamino-4-(isopropylamino)-6-(methythio)-s-triazine; desmetryne; G 34360; Topusyn; Methylamino-4-methylthio-6-isopropylamino-1,3,5-triazine; Topusyn; 2-(isopropylamino)-4-(methylamino)-6-(methyl-thio)-tri-azine;   N-methyl-N'-(1-methylethyl)-6-(methyl-thio)triazine; 2-Methylthio-4-methyl-amino-6-isopropyl-1,3,5-triazine. Selective systemic herbicide, inhibits photosynthesis. Post-emergence control of broad-leaf weeds and some grasses in brassicas,

herbs, onions, leeks, conifer seed beds. Wettable powder containing 25% w/w desmetryn; a triazine herbicide. mp = 84-86°; soluble in H_2O (580 mg/l), more soluble in organic solvents; LD_{50} (rat orl) = 1390 mg/kg.

830 Diallate

2303-16-4   3014   218-961-1

$C_{10}H_{17}Cl_2NOS$

S-(2,3-Dichloro-2-propenyl) bis(1-methylethyl)carbamothioate.

Avadex; DATC; Pyradex; DCDT. Herbicide for wild oats. *Monsanto Co; Monsanto plc.*

831 Dicamba

1918-00-9   3090   217-635-6

$C_8H_6Cl_2O_3$

3,6-Dichloro-2-methoxybenzoic acid.

3,6-dichloro-o-anisic acid; MDBA; Banvel; Fallowmaster; Mediben; Metambane; Tracker; Trooper; Velsicol 58-CS-11. Selective systemic herbicide which acts as an auxin-like growth regulator. For control of annual and perennial broadleaf weeds and brush species in cereals, maize, sorghum, sugar cane, asparagus, seed pastures, turf, pastures and rangeland. Herbicide used to control bracken. mp = 114-116°; d^{25} = 1.57; soluble in H_2O (6.5 g/l), more soluble in organic solvents; LD_{50} (rat orl) = 1707 mg/kg. *Diachem; Sandoz; Shell UK.*

832 Dichlobenil

1194-65-6   3093   214-787-5

$C_7H_3Cl_2N$

2,6-Dichlorobenzonitrile.

Prefix D; BH Prefix D; casoron; casoron-133; H-133; Niagara 5006; Carsoron; Casoron G; Casoron G4; Decabane; Prefix D; Casoron 10G; Casoron 4G; Casoron 50W; Dyclomec; Barrier; DCBN; Dichlorobenzonitrile; Du-sprex; Norosac; 2,6-Dichlorobenzoic acid nitrile; Fydulan; Fydumas; Fydusit. Granular herbicide containing dichlobenil; used to control weeds in woody crops and noncrop areas. Granules contain dichlobenil; residual herbicide for use among established trees and shrubs, paths, hard surfaces and vacant ground. Direct, selective weed killing action in orchards, vineyards, flower beds and parkland areas and along rail tracks, motorways and waterways. *Rhône-Poulenc Environmental Prods. Ltd.; Chipman Ltd.; ICI AgroChemicals; SynChemicals Ltd.; Duphar BV; Vitax Ltd.*

833 Dichloroprop

120 36 5   3128   204-390-5

$C_9H_8Cl_2O_3$

2-(2,4-Dichlorophenoxy)propanoic acid.

MSS 2,4-DP; 2,4-DP; α-(2,4-Dichlorophenoxy-propionic acid; Weedone 170; 2,4-D +

dichlorprop (ester); Weedone CB 1.3; BH 2,4-DP; 2,4-dichlorophenoxypropionic acid; Hedonal DP; Kildip; Polymone; Seritox 50; Weedone DP; Celatox DP; Cornox RK; Herbizid DP; Cornox RD; Desormone; DP; DP-fluid; Hormatox; Polytox; R-(+)-2-(2,4-dichlorophenoxy)propionic acid. Soluble concentrate of 500 g dichlorprop per liter; used for control of weeds in barley, wheat and oats. *Mirfield Sales Services Ltd.*

834 Diclofop-Methyl

51338-27-3   3133   257-141-8

$C_{16}H_{14}Cl_2O_4$

2-[4-(2,4-Dichlorophenoxy)-phenoxy]propanoic acid methyl ester.

Hoegrass; Diclofop-Methyl; HOE 23408; Heolon; Hoegrass; Hoelon; Hoelon 3EC; Iloxan; HOE-023408; Illoxan; Hoelon 3 EC; Hoelon 3 EW; Brestan H; Dichlordiphenprop; Diclofop methyl ester. Emulsifiable concentrate containing 378g diclofop-methyl per liter; used for control of weeds in grass. *Hoechst UK.*

835 Difenzoquat Methyl Sulfate

43222-48-6   3185   256-152-5

$C_{18}H_{20}N_2O_4S$

1,2-Dimethyl-3,5-diphenyl-1H-pyrazolium methyl sulfate.

Avenge 2; Avenge; Finaven; Yeh-yan-ku. Soluble concentrate of 150 g difenzoquat per liter; used for control of wild oats in cereals. *Cyanamid of Great Britain Ltd; Schering Agro-chemicals Ltd.*

836 Diflufenican

83164-33-4

$C_{19}H_{11}F_5N_2O_2$

N-(2,4-Difluorophenyl)-2-[3-(trifluoromethyl)-phenoxy]-3-pyridinecarboxamide.

diflufenicanil; M&B 38544; Kwarc; DFF. Selective contact and residual herbicide for control of broadleaf weeds and some grasses, particulalry *Galium, Veronica,* and *Viola* species in cereal crops. mp = 161-162°; almost insoluble in H_2O (0.05 mg/l), readily soluble in organic solvents; LD_{50} (rat orl) >2000 mg/kg.

837 Dimefuron

34205-21-5   251-879-4

$C_{15}H_{19}ClN_4O_3$

N'-(3-Chloro-4-(5-(1,1-dimethylethyl)-2-oxo-1,3,4-oxadiazol-3(2H)-yl)phenyl)-N,N-dimethylurea.

Pradone; Pradone Plus. Herbicide. *May & Baker Ltd.*

838 Dinitramine

29091-05-2   249-419-2

$C_{11}H_{13}F_3N_4O_4$

N^3,N^3-Diethyl-2,4-dinitro-6-(trifluoromethyl)-m-phenylenediamine.

Cobex; Usb 3584. Weedkiller containing dinitramine. *ICI Chemical & Polymers Ltd.*

839   Dinoseb
88-85-7   3341   201-861-7
$C_{10}H_{12}N_2O_5$
4,6-Dinitro-2-sec-butylphenol.
Dynamyte; Premerge; DNOSBP; F-ISO; Caldon; Vertac General and Selective Weed Killer; Basanite; Chemox General & PE; Chemsect; Dinitrax; Dinitro-3; Dinitro General; Drexel Dynamite 3; Dynamite; Elgetol 318; Hel-Fire; Kiloseb; Nitropone C; Subitex; Unicrop DNBP; Vertac Dinitro Weed Killer 5; Dynanap; Premerge Plus, with Dinitro; Klean Krop; DNBP; WSX-8365; Chemox PE; Dow General; Premerge; Dinitrobutyl phenol; Dinoseb. Herbicide for use on beans, small grains, forage, cereal crops. Used for the control of broadleaf weeds in peas, soybeans, potatoes, and orchards. *Draxel Chemical Company; Dow UK.*

840   Diphenamid
957-51-7   3364   213-482-4
$C_{16}H_{17}NO$
N,N-Dimethyl-α-phenylbenzeneacetamide.
Enide; Dymid; Enide 90W; Rideon; Dyfen; Diherbid; DIF 4; Dimid; ENIDE 50; Enide dinitro EC-tuco; Fenam; Lilly 34,314. Used for weed control in horticultural crops. *ICI Agrochemicals.*

841   Diquat dibromide
85-00-7   3415   201-579-4
$C_{12}H_{12}Br_2N_2$
1,1'-Ethylene-2,2'-dipyridylium dibromide.
FB/2; Aquacide; Reglone; Katalon; Midstream; Power Diquat; Reglox. A granular contact herbicide and pre-harvest crop desiccant and defoliant. A contact herbicide and pre-harvest crop desiccant. mp < 320° dec; also reported as mp = 335-340°; λ_m = 308.31 nm (ε 18000). *Makhteshim Chemical Works Ltd.; ICI Chem & Polymers Ltd.; Kommer-Brookwick Ltd.*

842   Diuron
330-54-1   3447   206-354-4
$C_9H_{10}Cl_2N_2O$
N'-(3,4-Dichlorophenyl)-N,N-dimethylurea.
1,1-dimethyl-3-(3,4-dichlorophenyl)urea; Diurex; Urox D.; Karmex; Karmex®; Direx® 4L; Diurex; Diuron; Diuron Bayer; Urox D; ; DMU; DCMU; Diurex; Aguron; M Velpar; Karmex; Urox D; Direx 4L; Direx 80W; Diuron 4L; Diuron 80; Karmex 80W; Karmex DL; Cekiuron; Crisuron; Dailon; Dion; Diater; Unidron; Vonduron; Xarmex, Krovar; Drexel diuron 4L; Dynex. Diuron; residual urea herbicide. A flowable herbicide for control of many weeds and grasses in a variety of crops. Effective against a wide range of both broadleaf weeds and annual grasses.   Pre-emergent

herbicide, sugar cane flowering suppressant. Inhibits photosynthesis. Used for total control of weeds and mosses in non-crop areas and selectiver control of germinating grass and broad-leaved weeds in many crops. mp = 158-159°. *Rohm & Haas UK; DuPont UK; Griffin; Agan Chemical Manufacturers Ltd.; Pacific Anchor; Rhône-Poulenc Agrochimie SA; Rhône-Poulenc Environmental Prods. Ltd.; Bayer AG.*

843   Dowpon
75-99-0   2869   204-828-5
$C_3H_3Cl_2NaO_2$
2,2-Dichloropropanoic acid sodium salt.
Dalapon sodium salt; Dowpon; Radapon; Dichloropropionic acid, sodium salt; Basfapon B. Herbicide; for controlling grass species; used primarily in sugar cane, sugar beets, orchards and also in noncrop applications such as railroads and rubber plantations. Post-emergence systemic herbicide for control of grasses in annual and perennial crops, used on nonagricultural land, in ditches, and pastures. mp = 174-176°; soluble in H_2O (45 g/100 ml); LD_{50} (rat orl) = 9330 mg/kg. *Dow UK; BASF AG.* See dalapon sodium salt.

844   Duplosan® DP
28631-35-8   249-110-2
$C_{17}H_{24}Cl_2O_3$
Isooctyl 2-(2,4-dichlorophenoxy)propionate.
Dichlorprop-P; DP, isooctyl ester. Herbicide for control of broadleaf weeds in cereals. *BASF AG.* See dichlorprop-P.

845   EPTC
759-94-4   212-073-8
C9H19NOS
S-Ethyl dipropylcarbamothioate.
Alirox; Eptam; Genep; Niptan; R-1608; Witox. Selective systemic herbicide, used to control annual and perennial grasses in many crops. Emulsifiable concentrate of 720 g EPTC per liter; used for pre-planting weed control in potatoes. bp_{20} = 127°; d^{30} = 0.9546; n_D^{30} = 1.4750; soluble in H_2O (375 mg/l), more soluble in organc solvents; LD_{50} (rat orl) = 1630 mg/kg. *Farm Protection Ltd.*

846   Ethofumesate
26225-79-6   3788   247-525-3
$C_{13}H_{18}O_5S$
(±)-2-Ethoxy-2,3-dihydro-3,3-dimethyl-5-benzofuranylmethanesulfonate.
Norton; Nortran; Nortron; Tramat; Prograss; Nortranese; NC-8438; Ethofumesate; betanal Tandem. Emulsifiable concentrate of 200 g ethofumesate per liter; used for weed control in field crops. mp = 69-71°; soluble in H_2O (110 mg/l), more soluble in organic solvents; LD_{50} (rat orl) = 6400 mg/kg. *Schering Agrochemicals Ltd.*

847 Fenoxaprop-ethyl

66441-23-4   4024   266-362-9

$C_{18}H_{16}CINO_5$

Ethyl(2-(4-((6-chloro-2-benzoxazolyen)oxy)-penoxy)propanic acid,ethyl ester).

Cheetah R; Acclaim; fenoxaprop; Whip; Horizon; Option; Fenoxaprop. An emulsion containing 60 g fenoxaprop-ethyl per liter; used for grass weed control in wheat. *Hoechst UK.*

848 Flamprop-isopropyl

52756-22-6   258-154-1

$C_{19}H_{19}CIFNO_3$

Isopropyl-N-benzoyl-N-(3-chloro-4-fluorophenyl)-alanine.

flamprop-isopropyl; Flamprop-M-isopropyl; Suffix BW; WL 29762; Barnon; Flufenprop-isopropyl; Commando; Gunner; Power Flame; Power Flamprop. Herbicide for control of wild oats in cereal crops. *Quadrangle Agrochemicals; Shell UK; Kommer-Brookwick Ltd.; Kommer-Brookwick Ltd.*

849 Flamprop-M Isopropyl

63782-90-1

$C_{19}H_{19}CIFNO_3$

5-((2-Chloro-α,α,α-trifluoro-p-tolyl)oxy)-2-nitrobenzoate.

Commando; Tackle 2AS. Emulsifiable concentrate of 200 g flamprop-M-isopropyl per liter; used for control of wild oats in cereal crops. *Shell UK.*

850 Flamprop-methyl

37924-13-3   7300   253-718-3

$C_{14}H_{12}F_3NO_4S_2$

1,1,1-Trifluoro-N-[2-methyl-4-(phenylsulfonyl)phenyl]methanesulfonamide.

Lancer; Perfluidone; MBR-8251; Destun; 2-methyl-4-phenylsulfonyltrifluoromethane sulfonanilide. Herbicide for wild oat control. MW = 379.38; solid,crystals; mp = 142-144°; soluble in H_2O, Me_2CO, C_6H_6, dichloromethane, MeOH. *ICI Chem & Polymers Ltd.*

851 Fluazifop-butyl

69806-50-4   4152   274-125-6

$C_{19}H_{20}F_3NO_4$

2-[4-[[5-(Trifluoromethyl)-2-pyridinyl]oxy]-phenoxy]propanoic acid butyl ester.

haloxyfop; Fusilade; Gallant; PP-009; TF-1169; butyl 2-[4-(5-trifluoromethyl-2-pyridyloxy)-phenoxy]propionate. Emulsifiable concentrate of 125 g fluazifop-p-butyl per liter; used for grass weed control for broad-leaved crops. bp = 167°. *ICI Chem & Polymers Ltd.; Dow UK.*

852 Fluometuron

2164-17-2   4189   218-500-4

$C_{10}H_{11}F_3N_2O$

1,1-Dimethyl-3-(α,α,α-trifluoro-m-tolyl)urea.

Cottonex; Meturon® 4L. Fluometuron suspension; flowable herbicide controlling annual grasses and broadleaf weeds in cotton and sugarcane. Residual herbicide effective against a wide range of both annual broadleaf weeds and grasses. *Agan Chemical Manufacturers Ltd.; Griffin.*

853 Fluroxypyr

69377-81-7   4238

$C_7H_5Cl_2FN_2O_3$

[(4-Amino-3,5-dichloro-6-fluoro-2-pyridinyl)oxy]acetic acid.

EF-689; Dowco 433; Starane. Selective systemic herbicide. Used for post-emergence control of broad-leaved weeds such as *Galium aparine* and *Stella media* and some deep-rooted perennial weeds. mp = 232-233°; soluble in H_2O (91 mg/l), more soluble in organic solvents; LD_{50} (rat orl) = 2405 mg/kg.

854 Fluroxypyr 1-Methylheptyl Ester

81406-37-3   4238   279-752-9

$C_{15}H_{21}Cl_2FN_2O_3$

[(4-Amino-3,5-dichloro-6-fluoro-2-pyridinyl)oxy]acetic acid 1-methylheptyl ester.

fluroxypyr-meptyl; Dowco 433; Starane. Selective systemic herbicide absorbed by leaves and roots. Used for control of many broad-leaved weeds in wheat and barley. mp = 56-57°; insoluble in H_2O (0.9 mg/l), freely soluble in organic solvents; LD_{50} (rat orl)= 2405 mg/kg.

855 Glyphosate

1071-83-6   4522   213-997-4

$C_3H_8NO_5P$

N-(Phosphonomethyl)glycine.

Glyphogan; MON-0573; Bronco; Landmaster; Ranger; Pondmaster; Rattler 4AS; Roundup; Roundup 2.5; glialka; MON 2139; MON 6000; phosphonomethyliminoacetic acid; N-phosphomethylglycine; Sonic; Spasor; Sting; tumbleweed. Broad spectrum translocatable herbicide. Monoisopropylamine salt is in Roundup. Non-selective systemic herbicide. For control of a wide variety of annual, biennial and perennial grasses, sedges, broad-leaved weeds and woody shrubs. A nonresidual herbicide containing 360 g/liter glyphosate for the control of annual and perennial broad-leaved weeds and grasses; used for clearing ground prior to planting, weed control in all hard surfaces. *Agan Chemical Manufacturers Ltd.; Schering Agrochemicals Ltd.; Rhône-Poulenc Environmental Prods. Ltd.; Monsanto plc.*

856 Gramoxone

1910-42-5   7165   217-615-7

$C_{12}H_{14}Cl_2N_2$

1,1'-Dimethyl-4,4'-Bipyridinium dichloride.

Gramoxone X; Paraquat; Scythe; Speedway; N,N'-dimethyl-σ,σ'-bipyridylium dichloride; methyl viologen dichloride; Crisquat; Dexuron; Esgram; Gramuron; Ortho Paraquat CL; Para-col; Pillarxone; Tota-col; Toxer Total; Paraquat Cl; paraquat dichloride trihydrate; PP148; Cyclone; Dichloride salt of paraquat; Gramixel; Gramoxone; Gramoxone dichloride; Gramoxone S; Gramoxone W; Pathclear; Methyl Viologen hydrate. Herbicide. Paraquat weedkiller preparations. Soluble concentrate containing 200 g/l paraquat; a pre-emergence bipyridinium herbicide to control weeds in field crops and ornamentals. *Schering Agrochemicals Ltd.; ICI Chem & Polymers Ltd.; ICI Agrochemicals; Schering Agriculture; Cyanamid of Great Britain Ltd.; Ashlade Formulations Ltd. See* paraquat.

857   Green-up Mossfree
7720-78-7   4105   231-753-5
FeO$_4$S
Ferrous sulfate.
Iron Vitriol; Duroferon; Iron sulfate; Iron monosulfate; Fer-In-Sol; Feosol. For moss control in turf. *Synchemicals Ltd. See* ferrous sulfate.

858   Hexazinone
51235-04-2   4734   257-074-4
C$_{12}$H$_{20}$N$_4$O$_2$
3-Cyclohexy-6-(dimethylamino)-1-methyl-1,3,5-triazine-2,4(1H,3H)-dione.
3-Cyclohexyl-6-(dimethylamino)-1-methyl-1,3,5-triazine-2,4(1H,3H)-dione; DPX 3674; Velpar; Hexazinone. Non-selective contact herbicide. Inhibits photosynthesis. Used for control of annual, biennial, perennial weeds and woody plants in non-crop areas and coniferous plantations. Soluble concentrate of 240 g hexazinone per liter; used for control of weeds in forestry plantations. mp = 115-117°; d = 1.25; soluble in H$_2$O (33 g/l), more soluble in organic solvents; LD$_{50}$ (rat orl) = 1590 mg/kg. *Selectokil Ltd (Velpar); DuPont UK (Velpar®).*

859   Imazapyr
81334-34-1   4942
C$_{13}$H$_{15}$N$_3$O$_3$
2-[4,5-Dihydro-4-methyl-4-(1-methylethyl)-5-oxo-1H-imidazol-2-yl]-3-pyridinecarboxylic acid.
AC 252,925; Arsenal; Assault; Chopper; CL 252,925; Contain; Pivot. Imazapyr with isopropylamine (1:1) salt; used for bracken control in noncrop areas. Non-selective systemic herbicide. Inhibits acetohydroxy acid synthase. Used for control of annual and perennial grass and broad-leaved weeds in non-crop areas. mp = 169-173°; soluble in H$_2$O (10-15 g/l), organic solvents; LD$_{50}$ (rat orl) >5000 mg/kg. *Chipman Ltd.; Am. Cyanamid/Ag; Cyanamid of Great Britain Ltd.; DowElanco Ltd.*

860   Isoproturon
34123-59-6   5237   251-835-4
C$_{12}$H$_{18}$N$_2$O
N,N-Dimethyl-N'-(4-(1-methylethyl)phenyl)urea.
Arelon; Chiltern IPU; Hytane; Portman Isotop; Power Swing; Protugan; Sabre; Tolkan; Tolkan 500; Alon; Belgran; Graminon; IP50; DPX 6774. Used for annual weed control in cereals. *Hoechst UK; Chiltern Farm Chemicals Ltd.; Ciba-Geigy Agrochemicals; Portman Agrochemicals Ltd.; Kommer-Brookwick Ltd.; Agan Chemical Manufacturers Ltd.; Schering Agrochemicals Ltd.; May & Baker Ltd.; Rhône-Poulenc Crop Protection Ltd.; Farmers Crop Chemicals Ltd.*

861   Isoxaben
82558-50-7   5256
C$_{18}$H$_{24}$N$_2$O$_4$
N-[3-(1-Ethyl-1-methylpropyl)-5-isoxazolyl]-2,6-dimethoxybenzamide.
N-[3-(1-Ethyl-1-methylpropyl)-5-isoxazolyl]-2,6-dimethoxybenzamide; benzamizole; EL-107; NA-8318; Flexidor; Gallery; Ratio; Tripart® Ratio; Knot Out. Suspension concentrate containing 125 g isoxaben per liter; used for control of annual dicotyledons in cereals, grass and fruit. A residual herbicide for the control of broad-leaved weeds in winter and spring cereals, grass leys and herbage seed crops. mp = 176-179°; poorly soluble in H$_2$O (1 mg/l), more soluble in organic solvents; LD$_{50}$ (rat orl) > 10000 mg/kg. *Synchemicals Ltd.; Tripart Farm Chemicals Ltd.*

862   Krenite
25954-13-6   247-363-3
C$_3$H$_{11}$N$_2$O$_4$P
Ammonium ethyl carbamoylphosphonate.
Krenite S; Krenite UT; DPX 1108; Phosphonic acid, (aminocarbonyl)-, monoethyl ester, monoammonium salt; Fosamine Ammonium Salt. Soluble concentrate of 480 g fosamine-ammonium per liter; used for control of woody weeds in noncrop and forestry areas. Applied to unwanted brush in late summer or autumn prevents bud break leading to death of treated plants the following spring. *Selectokil Ltd; Burts & Harvey; DuPont UK.*

863   Lenacil
2164-08-1   5459   218-499-0
C$_{13}$H$_{18}$N$_2$O$_2$
3-Cyclohexyl-6,7-dihydro-1H-cyclopentapyrimidine-2,4(3H,5H)-dione.
3-cyclohexyl-5,6-tri-methyleneuracil; 3-cyclo-hexyl-1,5,6,7-tetrahydo-2H-cyclopentapyrimidine-2,4(3H)-dione; Lenacil; du Pont 634; lenacil; 3-cyclohexyl-5,6-trimethyleneuracil; lenacile; Adol; Du Pont 634; Elbatan; Venzar. Selective herbicide, absorbed by roots, inhibits photosynthesis. Used for control of annual grass and broadleaf weeds in

a variety of crops. Wettable powder containing 80% lenacil for control of annual dicoyledons and meadow grass in beet, fruit, herbaceous perennials mp = 316-317°. *DuPont UK; ICI Chem & Polymers Ltd; Chemolimpex; Fahlberg-List; Du Pont; Farm Protection Ltd.*

864 Linuron

330-55-2   5534   206-356-5

$C_9H_{10}Cl_2N_2O_2$

N'-(3,4-Dichlorophenyl)-N-methoxy-N-methylurea.

Afalon; Ashlade Linuron; Atlas Linuron; Campbell's Linuron 45%; Du Pont Linuron 50, 4L; Rotalin; Methoxydiuron; Arresin. A residual urea herbicide for the control of weeds in field crops including potatoes and carrots. *Hoechst UK; Ashlade Formulations Ltd.; Atlas Interlates Ltd.; MTM AgroChemicals Ltd.; DuPont UK; Agan Chemical Manufacturers Ltd.; ICI Chem & Polymers Ltd.; Farm Protection Ltd.*

865 Ioxynil

1689-83-4   216-881-1

$C_7H_3I_2NO$

3,5-Diiodo-4-hydroxybenzonitrile.

Actrilawn 10; Actril; Certrol; Bantrol; CA 69-15; 4-Cyano-2,6-diiodophenol;   2,6-Diiodo-4-cyanophenol; Iotril; MATE; Trevespan; Totril. Contact herbicide for use in turf, onion crops. *Rhône-Poulenc Environmental Prods. Ltd.; Chipman Ltd.; ICI AgroChemicals; SynChemicals Ltd.; Duphar BV; Vitax Ltd.; Embetec Crop Protection Ltd.; May & Baker Ltd.*

866 Maloran

13360-45-7   236-411-9

$C_9H_{10}BrClN_2O_2$

N'-(4-Bromo-3-chlorophenyl)-N-methoxy-N-methylurea.

Maloran; chlorbromuron. Substituted urea herbicide. *Ciba plc. See* iron.

867 Maxicrop Moss Killer & Conditioner

10028-22-5   4079   233-072-9

$Fe_2O_{12}S_3$

Ferric sulfate.

Iron (III) sulfate; ferric persulfate; ferric sesquisulfate; ferric tersulfate; Sulfuric Acid, Iron(3+) Salt (3:2); Iron Tersulfate; Ferric sulfate monohydrate; Greenmaster Mosskiller; Vitax Micro Gran; Vitax Turf Tonic; Walkover Moss Killer; Elliott's Lawn Sand; Elliott's Moss Killer; Hart Lawn Sand; Hart Mosskiller; Greenmaster Autumn. Used for moss control in turf. *Maxicrop International Ltd.; Fisons plc; Vitax Ltd.; Walkover Sprayers Ltd.; Thomas Elliott Ltd.; Maxwell Hart Ltd.; Bayer AG. See* ferric sulfate.

868 MCPA

94-74-6   5803   202-360-6

$C_9H_9ClO_3$

(4-Chloro-2-methylphenoxy)acetic acid.

Farmon MCPA 50; Hedarex M; Hedonal; Hedonal M; Herbicide M; Hormotuho; Kilsem; Krezone; Leuna M; Linormone; Phenoxylene plus; Rhomene; Selektonon M; Shamrox; Vacate; Weed-rhap; Weedar; Weedone; Zelan; FBC MCPA; Agrichem MCPA-25, 50; Agricorn 500; Agritox 50; Agroxone 50; Albar-M; Atlas MCPA; BH MCPA 75; Chafer MCPA 675; Empal; Mecpa; MSS MCPA 50; Phenoxylene 50; Phenoxylene Plus; Power MCPA; Quad MCPA 50%; Star MCPA; Anicon m; BH MCPA; Cekherbex; Chloro-o-cresoxyacetic acid; Chloro-o-tolyloxyacetic acid; Chwastox; CMP acetate; Cornox-m; Dedweed; Dicopur-M; Dikotex; Emcepan; Empal; FLUID 4; Hedapur M 52; 2,4-MCPA; MCP; 4-Chloro-o-cresoxyacetic acid; Methyl chlorophenoxy acetic acid; MCPA Ester; Bordermaster; Metaxon; MCP ester; Weedar MCPA; Weedone MCPA Ester; Mephanac; Chiptox; Agroxon; Netazol; Rhonox; Anicon kombi. Selective, systemic, hormone-like herbicide used for post-emergence control of annual and perennial broad-leaved weeds in cereal crops and grassland. A herbicide used for cereals and grassland. mp = 118-119°; soluble in H_2O (825 mg/l), more soluble in organic solvents; LD_{50} (rat orl) = 700 mg/kg. *Farm Protection Ltd.; Schering Agrochemicals Ltd.; Agrichem (Int'l) Ltd.; Farmers Crop Chemicals Ltd.; Rhône-Poulenc Crop Protection Ltd.; ICI Agrochemicals; Makhteshim Chemical Works Ltd.; Atlas Interlates Ltd.; Rhône-Poulenc Environmental Prods. Ltd.; BritAg Industries Ltd.; Universal Crop Protection Ltd.; Murphy Chemical Co Ltd.; Mirfield Sales Services Ltd.; Schering Agrochemicals Ltd.; Fisons plc, Horticulture Div.; Kommer-Brookwick Ltd.; Quadrangle Agrochemicals; Star Agrochem Ltd.*

869 MCPB

94-81-5   202-365-3

$C_{11}H_{13}ClO_3$

4-(4-Chloro-2-methylphenoxy) butyric acid.

Belmac Straight; Fisons 18-15, MCPB; Tropotox; Can-Trol; PDQ; 2,4-MCPB; Bexane; MCP-butyric; Trifolex. Selective systemic hormone-like herbicide. Used for post-emergence control of annual and perennial broad-leaved weeds in cereal and grassland. Soluble concentrate containing 400 g/l MCPB; for control of weeds in undersown cereals and grassland. mp = 99-100°; soluble in H_2O (44 mg/l), more soluble in organic solvents; LD_{50} (rat orl) = 680 mg/kg. *MTM Agrochemicals Ltd.; Fisons plc, Horticultural Div.*

870   Mecoprop

7085-19-0   5826   230-386-8

$C_{10}H_{11}ClO_3$

(±)-2-(4-Chloro-2-methylphenoxy)propanoic acid.
Chafer CMPP Super; Cleanacres CMPP; Clenecorn; Clifton CMPP Amine 60; Clovotox; Compitox Extra; FBC CMPP; Headland Charge; Hymec; Iso-Cornox; Mascot Cloverkiller;   ; Clonotox; Compitox; Iso-cornox; Kilprop; Mecomec; Mecopex; Mepro; Propal; RD 4593; U 46 KV Fluid. Selective, systemic hormone-type herbicide, absorbed by leaves, translocated to roots. Post emergence control of broad-leaf weeds such as clovers, chickweeds, plantains and cleavers. Herbicide used for control of weeds in cereals and grassland. mp = 94-95°; soluble in H_2O (620 mg/l), soluble in organic solvents; LD_{50} (rat orl) = 930-1166 mg/kg. BritAg Industries Ltd.; Cleanacres Ltd.; Farmers Crop Chemicals Ltd.; Clifton Chemicals Ltd.; Rhône-Poulenc Environmental Prods. Ltd.; Rhône-Poulenc Crop Protection Ltd.; Schering Agrochemicals Ltd.; SBC Technology Ltd.; Agrichem (International) Ltd.; Schering Agrochemicals Ltd.; Rigby Taylor Ltd.; Rhône-Poulenc; Schering; BASF; Akzo; BASF; Fermenta; Marks; Rhône Poulenc; Universal Crop Protection; Bayer AG.

871   Mecoprop-P

93-65-2   202-264-4

2-(2-Methyl-4-chlorophenoxy) propionic acid.
Astix; Duplosan New System CMPP; Duplosan® CMPP; mecoprop-P; MCPP; Mecoprop; mechlorprop; CMPP; RD 4593; Astix CMPP; Lescopex 2.5L;; Methoxone M; Gordon's Mecomec. Selective herbicide for control of broadleaf weeds in cereals, meadows, pastures Soluble concentrate containing 600 g/l mecoprop-P (MCPP); used for control of weeds in undersown cereals and grassland. Rhône Poulenc Crop Protection Ltd.; BASF AG. See MCPP.

872   Mefenacet

73250-68-7   277-328-8

$C_{16}H_{14}N_2O_2S$

2-(2-Benzothiazolyloxy)-N-methyl-N-phenylacetamide.
Hinochloa®; FOE 1976; Hinochloa; NTN 801; Rancho. Herbicide effective against grasses (especially against Echinochloa crus-galli and some broad-leaved weeds in transplanted paddy rice; mainly used in combinations with other compounds. Selective herbicide, inhibits cell division. Used for control of grass weeds, especially Echinochloa cur-galli and cyperaceous weeds, pre- and early post-emergence in rice. mp = 135°; soluble in H_2O (4 mg/l), more soluble in organic solvents; LD_{50} (rat orl) > 5000 mg/kg. Bayer AG. See mefenacet.

873   Mefluidide

53780-34-0   5846   258-767-4

$C_{11}H_{13}F_3N_2O_3S$

N-[2,4-Dimethyl-5-[[(trifluoromethyl)sulfonyl]-amino]phenyl]acetamide.
Echo; Embark; MBR12325; Mowchem; Trimcut. A grass growth regulator containing 240 g/l mefluidide; suppresses most grasses for up to 8 weeks; for grassed areas not subject to heavy wear. Plant growth regulator and herbicide which inhibits growth and development of grasses. Used in lieu of grass cutting, e.g. on road verges and embankments. mp = 183-185°; soluble in H_2O (180 mg/l), more soluble in organic solvents; LD_{50} (rat orl) > 4000 mg/kg. Rhône-Poulenc Environmental Prods. Ltd.

874   Mephosfolan

950-10-7   5900   213-447-3

$C_8H_{16}NO_3PS_2$

(4-Methyl-1,3-dithiolan-2-ylidene)phosphoramidic acid diethyl ester.
Cytro-Lane; Cytrolane. Emulsifiable concentrate containing 250 g/l mephosfolan; used for control of damsonhop aphid in hops. Cyanamid of Great Britain Ltd. See mephosfolan.

875   Merpectogel

62-38-4   7453   200-532-5

$C_8H_8HgO_2$

Phenylmercuric acetate.
PMA; (acetato-O)phenylmercury; Acetoxyphenyl-mercury; (Aceto)phenylmercury; PMAC; PMAS; Gallotox; Liquiphene; Phix; Mersolite; Tag HL-331; Nylmerate; Scutl; Riogen; Advacide PMA 18; Cosan PMA; Mergal A25; Metasol 30; Nildew AC 30; Nuodex PMA 18; Agrosan; Cekusil; Celmer; Hong Nien; Pamisan; Seedtox; Shimmer-ex; Unisan; (acetoxymercuri)benzene; agrosan gn 5; algimycin; antimucin wdr; benzene, (acetoxymercurio)-; Bufen; ceresan universal; contra creme; dyanacide; Femma; FMA; fungitox or; HL-331; hostaquik; kwiksan; leytosan; Neantina; mersolite 8; norforms; phenmad; phenomercuric acetate; phenylmercuriacetate; pmacetate; PMAL; purasan-sc-10; puraturf 10; quicksan; quicksan 20; sanitized spg; SC-110; spor-kil; TAG; tag 331; trigosan; ziarnik; Anticon; Fungicide R; Fungitox; Meracen; Mercuron. Used as a herbicide and fungicide. Poythress Laboratories Inc. See phenylmercuric acetate.

876   Metamitron

41394-05-2   5985   255-349-3

$C_{10}H_{10}N_4O$

4-Amino-3-methyl-6-phenyl-1,2,4-triazin-5(4H)-one.
Goltix®; Countdown; BAY DRW 1139. A water dispersible granular formulation containing 70% w/w metamitron; a triazinone herbicide used to

control annual weeds in sugar beet grown on mineral and organic soils and red beet, fodder beet and mangolds grown on mineral soils. mp = 169°; LD$_{50}$(rat, orl) = 3343 mg/kg. *Bayer AG; Bayer plc; Kommer-Brookwick Ltd.*

877 Methabenzthiazuron
18691-97-9   6002   242-505-0
C$_{10}$H$_{11}$N$_3$OS
(Benzothiazol-2-yl)-1,3-dimethylurea.
Tribunil®; N-2-benzothiazolyl-N,N'-dimethylurea; Benzothiazolyl-1,3-dimethylurea; Benzothiazolyl-N,N'-dimethylurea; Dimethyl-3-(2-benzothiazolyl)urea; Dimethyl-3-(2-benzthiazolyl)-harnstoff; Methyl-N'-methyl-N'-(2-benzothiazolyl)urea; Preparation 5633; 1-(2-benzothiazolyl)-1,3-dimethyl-urea; Bay 72483; Methibenzuron. Selective, broadspectrum herbicide for pre- and post-emergence application. Photosynthesis inhibitor for control of annual grasses and broad-leaved weeds in garlic, onions, chives, leeks, peas, field beans, cereals, grass seed crops, lucerne, maize, potatoes, artichokes, stonefruit and tree nurseries. mp = 119-121°; soluble in H$_2$O (59 mg/l), more soluble in organic solvents; LD$_{50}$ (rat orl) > 2500 mg/kg. *Bayer AG; Bayer plc.*

878 Methar 30
144-21-8   6020   205-620-7
CH$_3$AsNa$_2$O$_3$
Methanearsonic acid disodium salt.
Methanearsonic Acid Na; Disodium methanearsonate; Ansar 8100; DMSA; Arrhenal; Arsinyl; Crab-E-Rad; Di-Tac; Methar 30; Sodar; Weed-E-Rad 360; Scotts Clout; Vertac dsma-lq; Weedone Crabgrass Killer; disodium methylarsonate; ; Methar; Ansar 184; Calar-e-rad; Chipco crab-kleen; DAL-E-RAD 100; Diarsen; Dinate; Disodium monomethylarsonate; Jon-trol; Maa, disodium salt; Metharsan; Metharsinat; Methylarsonic acid, disodium salt; Namate; Versar DSMA-LQ; Weed-e-rad. Selective herbicide, disodium methylarsonate (DSMA), for crabgrass control on grasses. *W A Cleary. See DSMA.*

879 Methazole
20354-26-1   6032   243-761-6
C$_9$H$_6$Cl$_2$N$_2$O$_3$
2-(3,4-Dichlorophenyl)-4-methyl-1,2,4-oxadiazolidine-3,5-dione.
oxydiazol; VCS-438; Paxilon; Probe; Probe 75 WP; Bioxone; Tunic; Chlormethazole; Mezopur. Broad spectrum herbicide. Wettable powder containing 75% methazole; for post-emergence weed control. mp − 123-124°; slightly soluble in H$_2$O (1.5 mg/l), more soluble in organic solvents; LD$_{50}$ (rat orl) = 777 mg/kg. *ICI Chem & Polymers Ltd.*

880 Metobromuron
3060-89-7   6224   221-301-5
C$_9$H$_{11}$BrN$_2$O$_2$
N'-(4-Bromophenyl)-N-methoxy-N-methylurea.
Ciba 3126; C 3126; Patoran; Patoran® FL; Pattonex. A substituted urea which inhibits photosynthesis; used for pre-emergence control of annual broad-leaved weeds and grasses in vegetable crops. For preemergence weed control in potatoes, soybeans, tobacco, tomatoes. mp = 95-96°; soluble in H$_2$O (330 mg/l), alcohols, more soluble in non-polar organic solvents; LD$_{50}$ (rat orl) = 3875 mg/kg. *Ciba plc; BASF AG; Agan Chemical Manufacturers Ltd.*

881 Metolachlor
51218-45-2   6230   257-060-8
C$_{15}$H$_{22}$ClNO$_2$
2-Chloro-N-(2-ethyl-6-methylphenyl)-N-(2-methoxy-1-methylethyl)acetamide.
Metolaclor; Dual; metelilachlor; CGA 24705; Bicep; Turbo; Bicep 6L; Dual 25G; Dual 8E; Pace 6L; Pennant; Primagram; Primextra; Codal; Ontrack 8E. Selective herbicide for control of annual grasses, some broad-leaved weeds. bp$_{0.001}$ = 100°; d = 1.12; soluble in H$_2$O (530 mg/l), more soluble in organic solvents; LD$_{50}$ (rat orl) = 2780 mg/kg.

882 Metoxuron
19937-59-8   243-433-2
C$_{10}$H$_{13}$ClN$_2$O$_2$
N'-(3-Chloro-4-methoxyphenyl)-N,N-dimethylurea.
Deftor; Dosanex; Dosaflo; Purivel; Sulerex. Suspension concentrate containing 500 g/l metoxuron; residual urea herbicide for the control of weeds in cereals and carrots. *ICI Chem. & Polymers Ltd; Farm Protection Ltd.*

883 Metribuzin
21087-64-9   6239   244-209-7
C$_8$H$_{14}$N$_4$OS
4-Amino-6-(1,1-dimethylethyl)-3-(methylthio)-1,2,4-triazin-5(4H)-one.
Sencor; Bayer 6159H; Lexone; 4-amino-6-tert-butyl-3-(methylthio)-as-triazin-5(4H)-one; Bay 94337; Sencorl;Preview; Salute; metribuzin + chlorimuron; Lexone 4L; Lexone 75DF; Lexone DF; Sencor 4L; Sencor 75DF; Sencor DF; Sencorex®; Sencor or metribuzin; Triazin-5(4H)-one, 4-amino-6-tert-butyl-3-(methylthio)-. Systemic herbicide for control of many grasses and broad-leaved weeds in soybeans, potatoes, tomatoes, sugarcane, alfalfa and asparagus; suitable for pre-and in some cases post emergence application. mp = 125°; SG = 1.28; d$_4^{20}$= 1.28;

slightly soluble in H_2O (1.2 g/l), soluble in MeOH, EtOH; LD_{50} (rat orl) = 2200 mg/kg. *Bayer plc.*

884   Metsulfuron-methyl
74223-64-6   6244
$C_{14}H_{15}N_5O_6S$
2-[[[[(4-Methoxy-6-methyl-1,3,5-triazin-2-yl)-amino]carbonyl]amino]sulfonyl] benzoic acid methylester.
Allie, Brushoff; Ally; DPX-6376; DPX-T6376; Escort; Granstar; Gropper. Selective systemic herbicide, used for pre- or post-emergence control of annual and perennial broad-leaved weeds in wheat, barley and oats. mp = 158°; soluble in H_2O (1.1 mg/l), more soluble in organic solvents; LD_{50} (rat orl) > 5000 mg/kg.

885   MSMA
2163-80-6   6020   218-495-9
CH_4AsNaO_3
Monosodium methylarsonate.
Ansar; Arsonate Liquid; Bueno; Daconate; Dal-E-Rad; Drexar; Drexar 530; Super Arsonate; Versar; Weed-E-Rad; Weed-Hoe; Neoarsycodyl; monosodium methylarsonate; monosodium methanearsonate. Used in post-emergence control of grass weeds in cotton, sugar cane and under trees, as a herbicide on turf to control established crabgrass, dollis grass, and nutselse. mp = 113-116°; soluble in H_2O (1.4 kg/kg), most organic solvents; LD_{50} (rat orl) = 900 mg/kg. *Lawn & Garden Products Inc; Drexel; Fermenta; Inter-Ag; Pamol; Shinung; Vertac; Vineland.*

886   Napropamide
15299-99-7   6503   239-333-3
$C_{17}H_{21}NO_2$
N,N-Diethyl-2-(1-naphthyloxy)propanamide.
Devrinol; Devrinol 10G; Devrinol 2E; Devrinol 50W; Waylay; Diethyl-2-(1-naphthyloxy)-propanamide; Napromide. Suspension concentrate containing 450 g/l napropamide; amide herbicide for oilseed, rape and fruit. *Embetec Crop Protection Ltd.*

887   Naptol
133-90-4   2115   205-123-5
$C_7H_5Cl_2NO_2$
2,5-Dichloro-3-aminobenzoic acid.
Chloramben; Ambiben; Amoben; Chlorambed; Ornamental; Weeder; Vegiben; ACP; M-629; Amiben; methyl; Weedone; Garden; Weeder; acp-m-728; amibin; amiben; ds; 3-Amino-2,5-dichlorobenzoic; acid. Chloramben; residual herbicide for use in ornamentals. *Synchemicals Ltd. See chloramben.*

888   Neburon
555-37-3   6523   209-096-0
$C_{12}H_{16}Cl_2N_2O$
N-Butyl-N'-(3,4-dichlorophenyl)-N-methylurea.
Granurex; Neburex; Noruben; Herbalt; 1-Butyl-3-(3,4-dichlorophenyl)-1-methylurea; Kloben; Butyl-3-(3,4-dichlorophenyl)-1-methylurea; Butyl-N'-(3,4-dichlorophenyl)-N-methylurea. Active ingredient: neburon; selective herbicide for both pre-and post-emergence application. Selective herbicide, absorbed through the roots, inhibits photosynthesis. Used for pre-emergence control of annual broad-leaved weeds and grasses in beans, peas, lucerne, garlic, cereals, beets, strawberries and ornamentals, and in forestry. mp = 102-103°; soluble in H_2O (5 mg/l), sparingly soluble in hydrocarbon solvents; LD_{50} (rat orl) > 11000 mg/kg. *Agan Chemical Manufacturers Ltd.*

889   Nitralin
4726-14-1   6662   225-219-0
$C_{13}H_{19}N_3O_6S$
4-(Methylsulfonyl)-2,6-dinitro-N,N-dipropylbenzeneamine.
SD 11831; Planavin. Herbicide. mp = 150-151°; slighlty soluble in H_2O (0.6 mg/l), more soluble in polar organic solvents; LD_{50} (rat orl) > 2 g/kg. *Shell Chemie GmbH.*

890   Oryzalin
19044-88-3   7015   242-777-0
$C_{12}H_8N_4O_6S$
4-(Dipropylamino)-3,5-dinitrobenzenesulfonamide.
3,5-dinitro-N^4, N^4-dipropylsulfanilamide; Dirimal; EL-119; Ryzelan; Surflan; Weed-Stopper. Selective herbicide; inhibits cell division and germination. Used for pre-emergence control of annual grasses. A pre-emergence herbicide for use on ornamentals, trees, roses, flower beds, bulbs, and warm season turf; controls annual grasses any many broadleaf weeds; may be tank-mixed with Roundup. mp = 141-142°; slightly soluble in H_2O (2.5 mg/l), more soluble in organic solvents; LD_{50} (rat orl) > 10000 mg/kg. *Lawn & Garden Products Inc.*

891   Oxadiazon
19666-30-9   7038   243-215-7
$C_{15}H_{18}Cl_2N_2O_3$
3-[2,4-Dichloro-5-(1-methylethoxy)phenyl]-5-(1,1-dimethylethyl)-1,3,4-oxadiazol-2(3H)-one.
Oxydiazon; Ronstar; Ronstar 2G; Ronstar 50W; Scotts OH I; RP-17623. A selective contact herbicide to control weeds and grasses in fruit and ornamental crops. mp = 88-89°; insoluble in H_2O (<1 mg/l), soluble in organic solvents; LD_{50} (rat orl) = 3500 mg/kg. *Embetec Crop Protection Ltd.*

892   Paraquat

4685-14-7   7165   225-141-7

$[C_{12}H_{14}N_2]^{2+}$

1,1'-Dimethyl-4,4'-bipyridinium.
Dextrone X; methyl viologen(2+); Paraquat + Plus. Non-selective contact herbicide. Soluble concentrate containing 200 g/l paraquat; a pre-emergence bipyridilium herbicide to control weeds in field crops and ornamentals. *ICI Agrochemicals Professional Products; Chipman Ltd.; Monsanto (Solaris).*

893   Pendimethalin

40487-42-1   7211   254-938-2

$C_{13}H_{19}N_3O_4$

N-(1-Ethylpropyl)-3,4-dimethyl-2,6-dinitrobenzenamine.
Penoxalin; Herbadox; Prowl; Pentagon; Stomp; Stomp H; Pre-M 60DG; Prowl 3.3E; Prowl 4E; Accotab; Herbodox; Go-Go-San; Way Up; Pay-off; Sipaxol. Selective herbicide, absorbed by roots and leaves. Used for control of most annual grasses and many broad-leaved weeds in cereal, vegetable and fruit crops. Emulsifiable or suspension concentrate containing pendimethalin; a dinitroaniline herbicide for cereals and bush fruit. mp = 56-57°; SG = 1.19; soluble in H_2O (0.3 mg/l), more soluble in organic solvents; LD_{50} (rat orl) = 1250 mg/kg. *Cyanamid of Great Britain Ltd.; Hortichem Ltd.*

894   Pentanochlor

2307-68-8   8851   218-988-9

$C_{13}H_{18}ClNO$

N-(3-Chloro-4-methylphenyl)-2-methylpentanamide.
3'-Chloro-2-methyl-o-toluidide; pentanochlore; CMMP; CMA; Croptex Bronze; FMC 4512; Dakuron; Solan; pentanochlor. Emulsifiable concentrate containing 400 g/l pentanochlor; used to control weeds in horticultural crops. mp = 85-86°; d_{20} = 1.106; soluble in H_2O (8-9 mg/l), freely soluble in organic solvents; LD_{50} (rat orl) > 10000 mg/kg. *Hortichem Ltd; Atlas-Interlates.*

895   Phenmedipham

13684-63-4   7384   237-199-0

$C_{16}H_{16}N_2O_4$

3-[(Methoxycarbonyl)amino]phenyl   (3-methylphenyl)carbamate.
Gusto; Headland Dephend; Beetomax; Beetup; Betalion; Betanal E; Betanal Tandem; Goliath; Pistol; Pistol 400; Protrum K; Suplex; Tripart® Beta, Tripart® Beta 2; Vangard; Vanguard; Schering 38584; Betanal. Selective systemic herbicide, absorbed through the leaves, inhibits photosynthesis. Used for post-emergence control of annual broad-leaved weeds in sugar beet, fodder beet, beetroot, mangels, spinach and strawberries. Selective systemic herbicide, absorbed through the leaves, inhibits photosynthesis. Used for post-emergence control of annual broad-leaved weeds in sugar beet, fodder beet, beetroot, mangels, spinach and strawberries. *SBC Technology Ltd; Fine Agrochemicals Ltd; MTM Agrochemicals Ltd; Schering Agrochemicals Ltd; ABM Chemicals Ltd; Farm Protection Ltd; Rhone-Poulenc UK; Atlas Interlates Ltd; Universal Crop Protection Ltd; Tripart Farm Chemicals Ltd;Farmers Crop Chem. Ltd.*

896   Planotox

1929-73-3   217-680-1

$C_{14}H_{18}Cl_2O_4$

(2,4-Dichlorophenoxy)acetic acid 2-butoxyethyl ester.
Planotox; 2,4-D, butoxyethanol ester; 2,4-D, 2-butoxyethyl ester; Aqua-kleen; BEE; Bladex-B; Brush killer 64; Butoxyethanol ester of 2,4-D; Weedone LV 4. Selective weedkiller. *May & Baker Ltd.*

897   Prochloraz

67747-09-5   7941   266-994-5

$C_{15}H_{16}Cl_3N_3O_2$

N-Propyl-N-(2-(2,4,6-trichlorophenoxy)ethyl)-1H-imidazole-1-carboxamide.
Sportak; Octave. Emulsifiable concentrate containing prochloraz; broad-spectrum fungicide for cereal crops. *Schering Agrochemicals Ltd.*

898   Prometryn

7287-19-6   7973   230-711-3

$C_{10}H_{19}N_5S$

N,N'-Bis(1-methylethyl)-6-(methylthio)-1,3,5-triazine-2,4-diamine.
Cotton-Pro®; s-triazine, 2,4-bis(isopropylamino)-6-methylthio-; s-triazine, 2,4-bis(isopropylamino)-6-methylmercapto-; A 1114; Caparol; G 34161; Gesagard; Gesagard 50; Gesagarde 50 Wp; Mercasin; Mercazin; Merkazin; Polisin; Primatol Q; Prometrex; Prometrex; Prometrin; Prometryne; Selectin; Selectin 50; Selektin; Sesagard; Uvon; 2-(methylmercapto)-4,6-bis(isopropylamino)-s-triazine; 2-(methylthio)-4,6-bis(isopropylamino)-s-triazine; Caparol; Gesagard; Caparol 4L; Caparol 80W; Cotton-Pro; Primatol Q; Prometrex; G-34161; Selectin; Uvon. A selective pre- and post-emergence herbicide for the control of broadleaf and grass weeds in a variety of crops. Absorbed by roots and foliage. Used for pre- and post-emergence control of most annual grasses and broad-leaved weeds. Used for selective weed control in cotton and celery crops. mp = 118-120°; d_{20} = 1.157; soluble in H_2O (48 mg/l), more soluble in organic solvents; LD_{50} (rat orl) = 1800 mg/kg. *Griffin; Ciba Geigy Agrochemicals; Agan Chemical Manufacturers Ltd.*

899   Propachlor

1918-16-7   7977   217-638-2

$C_{11}H_{14}ClNO$

2-Chloro-N-(1-methylethyl)-N-phenylacetamide.
Albras Propachlor; Albrass; Atlas Orange; Bexton; Croptex Amber; Portman propachlor 50FL; Prolex; Ramrod; Tripart® Granular; Tripart® Sentinel; Bexton 4L; Niticid; Satecid; Propachlor; 2-chloro-N-isopropylacetanilide;CP 31393; Bexton; Prolex;Propaclor; Ramrod-atrazine; Ramrod 20G; Ramrod Flowable; propachlor + atrazine; Chloro-N-isopropylacetanilide; Isopropyl-2-chloroacet-anilide; Albrass; Cp 31393; Croptex; Amber; Niticid; Orange; Prolex; Ramrod; Satecid; Sentinel. A pre-emergence herbicide for various horticultural crops. Selective herbicide, absorbed by seedling shoots and roots. Used for control of annual grasses and some broad-leaved weeds in vegetable crops. Herbicide used for pre-emergence grass and broadleaf weed control in corn and grain sorghum. mp = 77°; $bp_{0.03}$ = 110°; soluble in H_2O (613 mg/l), more soluble in organic solvents; LD_{50} (rat orl) = 1800 mg/kg. *ICI Chem & Polymers Ltd.; ICI Plant Protection; Atlas Interlates Ltd.; Dow UK; Hortichem Ltd.; Portman Agrochemicals Ltd.; Agan Chemical Manufacturers Ltd.; Monsanto plc; Tripart Farm Chemicals Ltd.*

900   Propanil

709-98-8   7987   211-914-6

$C_9H_9Cl_2NO$

N-(3,4-Dichlorophenyl)propanamide.
Propanil; Sorpur®; DPA; FW-734; Stam; Stampede; Rogue; Chem Rice; Surcopur; Bay 30130; Cekupropanil; Erbanil; Herbax; Prop Job; Propa; Propal; Propanex; Prostar; Riselect; Strel; Supernox; Surpur; Wham. Herbicide and nematocide. Post-emergence applied herbicide with no reidual effect. Used for control of numerous grasses and broad-leaved weeds in rice crops. mp = 91-93°; soluble in H_2O (225 mg/l); LD_{50} (rat orl) = 1384 mg/kg. *Bayer AG.*

901   Propazine

139-40-2   7996   205-359-9

$C_9H_{16}ClN_5$

6-Chloro-N,N'-bis(1-methylethyl)-1,3,5-triazine-2,4-diamine.
G-30028; Gesamil; Milogard; Prozinex. Pre-emergent selective systemic herbicide used for control of annual grasses and broad-leaved weeds in sorghum and crops such as carrots, chervil and parsley. Active ingredient; propazine; 2-chloro-4, 6-bis-(isopropylamino)-1,3,5-triazine; selective pre-emergent herbicide. mp = 213°; soluble in H_2O (8.6 mg/ml), poorly soluble in organic solvents; LD_{50} (rat orl) >7000 mg/kg. Agan Chemical Manufactures Ltd.

902   Propham

122-42-9   8001   204-542-0

$C_{10}H_{13}NO_2$

Phenylcarbamic acid 1-methylethyl ester.
MSS IPC 50; INPC; IPC; IsoPPC; Chem-Hoe; Chem Hoe FL4; Isopropyl carbanilate; Agermin; Birgin; Collavin; IFC; IFK; INPC; IPPC; Isoppc; Isopropyl phenyl urethane; Ortho grass killer; Phenyl isopropyl carbamate; Premalox; Profam; Tixit; Triherbide. Wettable powder containing 50% w/w propham; used for weed control for beet crops and peas. Propham; plant growth regulator for control of sprouting in stroed potatoes and in some cases as herbicide against weeds in vegetables. mp = 90°; insoluble in H_2O, soluble in organic solvents; LD_{50} (rat orl) = 3724 mg/kg. *Bayer AG.*

903   Propyzamide

23950-58-5   8058   245-951-4

$C_{12}H_{11}Cl2_NO$

3,5-Dichloro-N-(1,1-dimethyl-2-propynyl)benzamide.
Kerb; Kerb 50W; pronamid; RH-315; Kerb Propyzamide 50; Campbell's Rapier; Rapier. A residual herbicide in a wettable powder form for a wide range of agricultural crops such as oil seed rape. oil seed rape. mp = 155-156°; vapor pressure at 25° = 8.5×10^{-5} mm Hg. *Rohm & Haas UK; Pan Britannica Industries Ltd.; Kommer-Brookwick Ltd.; MTM AgroChemicals Ltd.; Farmers Crop Chemicals Ltd.*

904   Pyridate

55512-33-9

$C_{19}H_{23}ClN_2O_2S$

O-(6-Chloro-3-phenyl-4-pyridazinyl) S-octyl carbonothioate.
Tough; Lentagran; CL 11344; Fenpyrate; Screen; ST 9551. Wettable powder containing 45% w/w pyridate; for annual weed control for cereals, oilseed rape and maize. *Ciba Geigy Agrochemicals.*

905   Quinclorac

84087-01-4   402-780-1

$C_{10}H_5Cl_2NO_2$

3,7-Dichloro-8-quinolinecarboxylic acid.
Facet®; BAS-514-H; Facet; Facet 75 DF. For post-emergence control of barnyard grass and some other weeds in rice. mp = 274°. *BASF AG.*

906   Quizalofop-ethyl

76578-14-8   8269

$C_{19}H_{17}ClN_2O_4$

2-[4-[(6-Chloro-2-quinoxalinyl)oxy]phenoxy]propanoic acid ethyl ester.
quinofop-ethyl; DPX-Y6202; NCI-96683; NC-302; Assure; Targa; Pilot; EXP-3864; FBC-32197; INY-

6202. Post-emergence herbicide used for control of grasses in broad-leaved crops. Selective herbicide for control of grasses in mustard, rape and beet crops. mp = 92-93°; bp$_{0.2}$ = 220°; insoluble in H$_2$O, soluble in organic solvents; LD$_{50}$ (rat orl) = 1670 mg/kg. *Schering Agrochemicals Ltd.*

907   Rodeo
38641-94-0   254-056-8
C$_6$H$_{17}$N$_2$O$_5$P
Phosphonomethyl)glycine isopropylamine salt.
Roundup; Quick; Revoke; Glyphosate isopropylamine salt; Drat; Liphadione; Glyphosate Amine; Glifonox; Glycel; Glycine, N-(phosphonomethyl)-, compd. with 2-propanamine (1:1); Isopropylamine glyphosate; Phosphonomethyl)glycine, isopropylamine salt; Rodeo; Rondo. Aquatic herbicide. *Monsanto Co.*

908   Sethoxydim
74051-80-2   8620   277-682-3
C$_{17}$H$_{29}$NO$_3$S
(±)-(EZ)-2-(1-Ethoxyiminobutyl)-5-[2-(ethylthio)propyl]-3-hydroxycyclohex-2-enone.
sethoxydime; BAS 90520H; Checkmate; Expand; Fervinal; Grasidim; Nabu; NP-55; Poast; Poast®; SN 81742. Selective systemic herbicide absorbed by foliage, used in control of annual and perennial grasses in broad-leaved crops. Post-emergence graminicide against annual and perennial grasses. bp$_{0.00003}$ > 90°; d^{25} = 1.043; soluble in H$_2$O (25 mg/l), soluble in organic solvents; LD$_{50}$ (rat orl)= 3200 mg/kg. *BASF; Rhône-Poulenc; Ewos; Schering; Sipcam; Nippon Soda; Embetec Crop Protection Ltd.*

909   Silvex
93-72-1   8679   202-271-2
C$_9$H$_7$Cl$_3$O$_3$
2-(2,4,5-Trichlorophenoxy)propionic acid.
Kuron; fenoprop; 2,4,5-TC. Herbicide containing silvex as the active ingredient; herbicide used in ponds and other still water for the control of aquatic weeds, as well as control of brush on rangeland; also used industrially on railroads or under power lines for the control of weeds and brush. mp = 177-170°, also 181.6°. *Dow UK.*

910   Simazine
122-34-9   8681   204-535-2
C$_7$H$_{12}$ClN$_5$
6-Chloro-N,N'-diethyl-1,3,5-triazine-2,4-diamine.
2-chloro-4,6-bis(ethylamino)-s-triazine; 2,4-bis-(ethylamino)-6-chloro-s triazine; A 2079; Aktinit S; Amizine; Aquazine; Batazina; Bitemol; Bitemol S 50; Bitemol S-50; Cat (herbicide); Cekusan; Cekuzina-S; CAT; CAT (herbicide); CDT; CET; DCT; ENT-51142; Framed; G 27692; Geigy 27,692; Gesapun; Gesatop; Gesatop-50; H 1803;

Herbazin; Herbazin 50; Premazine; Primatol S; Princep; Printop; Radocon; Radokor; Weedex S2; Boroflow; Gesatop; Herbazin 50; Simadex; Simanex; Simapron; Simflow; Sinazine; Syngran. Selective systemic herbicide, absorbed through roots; used to control most germinating grasses and broad-leaved weeds in fruit, vegetables, trees Suspension concentrate containing 500 g/l simazine; a triazine herbicide to control weeds and grasses in cane fruit, roses and some vegetables. Long term maintenance of weed-free pathways, bare ground and other areas requiring total weed control. mp = 225-227sg; d$_{20}$ = 1.302; soluble in H$_2$O (5 mg/l), soluble in organic solvents; LD$_{50}$ (rat orl) = 971 mg/kg. *Ciba-Geigy Agrochemicals; Sedagril; Hoechst; Pepro; Protex; Sipcam-Phytoeurop.; Griffin; Schering; Schering Agrochemicals Ltd.; Rhône-Poulenc Environmental Prods. Ltd.; Hortichem Ltd.; ABM Chemicals Ltd.; Fisons plc, Horticultural Div.; Agan Chemical Manufacturers Ltd.; BP Oil Ltd.; Murphy Chem Co Ltd.; Synchemicals Ltd.; Mirfield Sales Services Ltd.*

911   Sodium Chlorate
7775-09-9   8741   231-887-4
ClNaO3
Chloric acid sodium salt.
Arpal Non Selex; Atlacide; Granular Weedkiller; Centex; soda chlorate; chlorate of soda; Drop-Leaf; Fall; Harvest-Aid; Tumbleaf; Altacide;Chlorax; Shed-A-Leaf 'L'. Powder containing 58.2% w/w sodium chlorate; used for total weed control for paths, drives and noncrop areas. *R. P. Adams Ltd.; Chipman Ltd.; Dimex Ltd.; Chemsearch (UK) Ltd.*

912   Sodium Metaborate
7775-19-1   8785   231-891-6
NaBO$_2$
Monosodium metaborate.
boric acid (HBO$_2$), sodium salt; sodium borate (NaBO$_2$); sodium metaborate (NaBO$_2$). Herbicide. mp = 966°; soluble in H$_2$O. *Ashland; U.S. Borax & Chem.*

913   Sodium Monochloracetate
3926-62-3   2162   223-498-3
C$_2$H$_2$ClNaO$_2$
Chloroacetic acid sodium salt.
monoxone; sodium chloroacetate; sodium mono-chloroacetate; SMA; SMCA. Herbicide. Soluble in H$_2$O (85 g/100 ml); LD$_{50}$ (rat orl) = 76 mg/kg.

914   2,4,5-T
93-76-5   9194   202-273-3
C$_8$H$_5$Cl$_3$O$_3$
(2,4,5-Trichlorophenoxy)acetic acid.
Trioxone; Esterone 245; Trioxone; Dacamine 4T; Farmco Fence Rider; Forron; Inverton 245; Line

Rider; Super D Weedone; Tributon; U 46; bcf-bushkiller; brush-off 445 low volatile brush killer; brush rhap; brushtox; decamine 4t; ded-weed brush killer; ded-weed lv-6 brush kil and t-5 brush kil; envert-t; estericide t-2 and t-245; esteron 245 be; esteron brush killer; fence rider; forst u 46; fortex; fruitone a; phortox; reddon; reddox; spontox; Super; tippon; VEON; veon 245; verton 2t; visko rhap low volatile ester; weedone 2,4,5-t; Dacamine; Dinoxol; Esteron; Tormona; Transamine; Trinoxol. Herbicide. *ICI Chem.*

915   Tandem
58138-08-2   9795
$C_{10}H_7Cl_5O$
(±)-2-(3,5-Dichlorophenyl)-2-(2,2,2-trichloroethyl)oxirane.
(RS)-2-(3,5-dichlorophenyl)-2-(2,2,2-trichloro-ethyl)oxirane; Dowco 356; Nelpon. A line of herbicides based primarily on tridiphane, a selective non-systemic herbicide. Used for control of annual grass seedlings and broad-leaved weeds in maize. mp = 43°; soluble in H_2O (1.8 mg/l), very soluble in organic solvents; LD_{50} (rat orl) = 1743-1918 mg/kg. *Dow UK.*

916   Tebithiuron
34014-18-1   9255   251-793-7
$C_9H_{16}N_4OS$
N-[5-(1,1-Dimethylethyl)-1,3,4-thiadiazol-2-yl]-N,N'-dimethylurea.
Bushwacker; Graslan; Spike; Perflan; Graslan 40P; Spike 20P; Spike 40P; Spike 40W; Spike 5G; Spike 80W; Spike DF. Wettable powder or granules used for total weed control in noncrop areas. *Rhône-Poulenc Environmental Prods. Ltd.*

917   Tebutam
35256-85-0   252-470-3
$C_{15}H_{23}NO$
2,2-Dimethyl-N-(1-methylethyl)-N-(phenylmethyl)propanamide.
N-Benzyl-N-isopropylpivalamide; Comodor 600; tebutame; butam; Comodor; GPC-5544; Ro 14-9480/000. Selective herbicide, acting by inhibition of weed germination. Used for pre-emergence control of annual grasses and broad-leaved weeds. Emulsifiable concentrate containing 720 g/l or 600 g/l tebutam; weed germination inhibitor. BP;si0.1 = 95-97°; d_{25} = 0.975; soluble in H_2O (0.79 mg/l. 25°), readily soluble in organic solvents; LD_{50} (rat orl) = 6210 mg/kg. *Maag; ICI Farm Protection; La Quinoleine; Roche; Farm Protection Ltd.*

918   Temephos
3383-96-8   9286   222-191-1
$C_{16}H_{20}O_6P_2S_3$
O,O'-(Thiodi-4,1-phenylene)phosphorothioic acid O,O,O',O'-tetramethyl ester.
Difenphos; Phosphorothioic acid, O,O'-(thiodi-p-phenylene) O,O,O',O'-tetramethyl ester; Abate 1-SG, 2-CG, 4-E, 5CG; Phosphorothioic acid, O,O'-(thiodi-p-phenylene) O,O,O',O'-tetramethyl ester. Granular and emulsifiable concentrated herbicide. *Amercian Cyanamid/Ag.*

919   Terbacil
5902-51-2   9298   227-595-1
$C_9H_{13}ClN_2O_2$
5-Chloro-3-(1,1-dimethylethyl)-6-methyl-2,4-(1H,3H)-pyrimidinedione.
Sinbar®; Sinbar 80W; DPX-D732; 3-tert-butyl-5-chloro-6-methyluracil; Geonter; butyl-5-chloro-6-methyluracil; chloro-3-*tert*-butyl-6-methyluracil; Herbicide 732; 5-chloro-3-(1,1-dimethylethyl)-6-methyl-pyrimidinedione,. Selective herbicide, used for control of annual broad-leaved weeds, most annual grasses and some perennial weeds. mp = 175-177°; SG - 1.34; soluble in H_2O (710 mg/l), more soluble in organic solvents; LD_{50} (rat orl) > 5000 mg/kg. *DuPont UK.*

920   Terbuthylazine
5915-41-3   227-637-9
$C_9H_{16}ClN_5$
2-t-Butylamino-4-chloro-6-ethylamino-1,3,5-triazine.
Tyllanex; Gardoprim; GS 13529; Primatol M. Herbicide, absorbed mainly by roots. Used for broad spectrum weed control. mp = 177-179°; SG_{20} = 1.188; soluble in H_2O (8.5 mg/l), more soluble in organic solvents; LD_{50} (rat orl) = 2160 mg/kg. *Agan Chemical Manufacturers Ltd.*

921   Timbrel
55335-06-3   9789   259-597-3
$C_7H_4Cl_3NO_3$
[(3,5,6-Trichloro-2-pyridinyl)oxy]acetic acid.
3,5,6-trichloro-2-pyridyloxyacetic acid; Ace-Brush; Crossbow; Dowco 233; Exetor; Garlon; Mutan; Redeem; Rely; Remedy; Turflon. Timbrel is an emulsifiable concentrate containing 480 g/l triclopyr; herbicide to control perennial and woody weeds. mp = 148-150°; bp= 290° (dec); soluble in H_2O (440 mg/l), more soluble in organic solvents; LD_{50} (rat orl)= 713 mg/kg. *DowElanco Ltd.*

922   Tordon
1918-02-1   7552   217-636-1
$C_6H_3Cl_3N_2O_2$
4-Amino-3,5,6-trichloro-2-pyridine carboxylic acid.
4-amino-3,5,6-trichloropicolinic acid; picloram; Grazon; Tordon 22K. Herbicides based primarily on picloram; broadleaf and brush killers for forestry, grain and corn. Soluble concentrate containing 240 g/l picloram; a picolinic herbicide to control woody weeds in noncrop areas. mp =

215° (dec); soluble in H_2O (430 mg/l), more soluble in organic solvents; LD_{50} (rat orl) = 8200 mg/kg. *Dow UK; Chipman Ltd.*

923 Triallate

2303-17-5   9726   218-962-7
$C_{10}H_{16}Cl_3NOS$
S-(2,3,3-Trichloro-2-propenyl)
bis(1-methylethyl)carbamothioate.
Avadex® BW; Far-Go/Avadex BW; Trichloroallyl diisopropylthiocarbamate; CP 23426; Avadex BW; Avadex BE; Far-Go; Showdown. Herbicide for control of wild oats, slender foxtail and bent grass in sugar beet and feed turnips, summer and winter barley, winter rye. mp = 29-30°; bp_{40} = 117°; d = 1.273; slightly soluble in H_2O (4 mg/l), more soluble in organic solvents; LD_{50} (rat orl) = 1100 mg/kg. *BASF AG; Monsanto Co.*

924 Trichloroacetic Acid

76-03-9   9756   200-927-2
$C_2HCl_3O_2$
Trichloroacetic acid.
TCA; NaTa; Farmon TCA. For control of weeds in field crops; also as a decalcifier and fixative in microscopy and a protein precipitant. mp = 54-58°; bp = 196-197°; d = 1.629; soluble in H_2O (10 g/ml), organic solvents; LD_{50} (rat orl) = 5000 mg/kg. *Farm Protection Ltd.; Hoechst UK.*

925 Triclopyr

69633-04-1
$C_8H_7Cl_3O_5S$
2-(2,4,5-Trichlorophenoxy)ethanol hydrogen sulfate.
Garlon; Garlon 2; Garlon 4. Emulsifiable concentrate containing 240 g/l triclopyr, herbicide to control perennial and woody weeds. *Dow Cheml Co Ltd, UK & Ireland; Dow UK; ICI Chem & Polymers Ltd.; Chipman Ltd.*

926 Trietazine

1912-26-1   9797   217-618-3
$C_9H_{16}ClN_5$
6-Chloro-N,N,N'-triethyl-1,3,5-triazine-2,4-diamine.
2-chloro-4-diethylamino-6-ethylamino-s-triazine; 2-chloro-4,6-diethylamino-s-triazine;   G-27901; NC-1667; Aventox; Gesafloc. Herbicide. mp = 100-102°; soluble in H_2O (20 mg/l), more soluble in organic solvents; LD_{50} (rat orl) = 1750 mg/kg. *Schering Agrochemicals Ltd.*

Herbicides containing 2,4-D

927 Agrichem DB Plus

2,4-DB + MCPA; translocatable herbicide for cereal crops. *Agrichem (International) Ltd.*

928 Alistell

Linuron + 2,4-DB-MCPA.
Herbicide containing linuron, 2,4-DB and MCPA. *ICI Chem & Polymers Ltd.*

929 Atladox HI

Soluble concentrate containing 240 g 2,4-D and 65 g picloram per liter; used to control weeds in non crop grass and grass verges. *Chipman Ltd. See* 2,4-D, picloram.

930 Bellclo

Soluble concentrate containing 250g 2,4-DB and 53 g mecoprop per liter; a translocated herbicide. *MTM Agrochemicals Ltd. See* 2,4-DB, mecoprop.

931 BH CMPP/2,4-D

2,4-D + Mecoprop.
Soluble concentrate containing 116g 2,4-D and 250g mecoprop per liter; used to control weeds in grassland. *Rhône-Poulenc/Agri.*

932 Broadshot

Emulsifiable concentrate containing 200 g 2,4-D, 85 g dicamba and 65 g triclopyr per liter; an herbicide to control perennial and woody weeds. *Shell UK. See* 2,4-D, dicamba, triclopyr.

933 Brush Buster

2,4-D and 2,4-DP; A post-emergence herbicide used to control woody species such as poison oak/ivy and brambles; applied as a foliar spray or straight from the bottle to the cut stump of woody species. *Lawn & Garden Products Inc.*

935 Campbell's Redlegor

2893
2,4-DB [94-82-6] + MCPA [94-74-6].
Translocated herbicide for cereal crops. *MTM AgroChemicals Ltd. See* 2,4-DB, MCPA.

936 Clovacorn Extra

An emulsifiable concentrate containing 220 g 2,4-DB, 30 g linuron and 30 g MCPA per liter; used to control weeds in undersown cereals and seedling grassland. *Farmers Crop Chemicals Ltd. See* 2,4-DB, linuron, MCPA.

937 Destral

2,4-D-Dalapon + diuron.
Mixture of 2,4-D, dalapon and diuron; used for total weed control in non crop areas. *ABM Chemicals Ltd. See* 2,4-D, dalapon, diuron.

938 Estermone

2,4-D + dicamba.
Mixture of 2,4-D and dicamba; used to control weeds in turf. *Synchemicals Ltd.*

939   Farmon 2,4-DB Plus
2,4-DB-MCPA.
2,4-DB and MCPA; a herbicide for use with cereal crops. *Farm Protection Ltd. See* 2,4-DB, MCPA.

940   Green Up
2,4-D + dicamba.
Mixture of 2,4-D and dicamba; used to control weeds in turf. *Synchemicals Ltd.*

941   Green Up Feed and Weed Plus Mosskiller
Dry powder containing NPK 8:4:4 plus 2,4-D, mecoprop and ferrous sulfate; combined fertilizer, weed and mosskiller. *Vitax Ltd.*

942   Green Up Lawn Feed and Weed
Liquid concentrate containing NPK 14.5:3:3 and 2,4-D and dicamba; combined feed and weed for turf. *Vitax Ltd.*

943   Green Up Lawn Spot Weedkiller
2,4-D + mecoprop.
Trigger spray pack containing 2,4-D [1702-17-6] and mecoprop [7085-19-0]; selective spot weedkiller for lawn areas. *Vitax Ltd. See* 2,4-D, mecoprop.

944   Green Up Weedfree Lawn Weedkiller
2,4-D + dicamba.
Emulsifiable concentrate containing 2,4-D [1702-17-6] and dicamba [1918-00-9]; selective herbicide for use on turf. *Vitax Ltd. See* 2,4-D, dicamba.

945   Green Up Weedfree Spot Weedkiller for Lawns
2,4-D + dicamba.
Aerosol containing 2,4-D [1702-17-6] and dicamba [1918-00-9]; selective spot weedkiller for lawns. *Vitax Ltd. See* 2,4-D, dicamba.

946   Hedonal®
2,4-D MCPA + Dichlorprop-MCPP.
Range of herbicides containing 2,4-D MCPA, dichlorprop, MCPP either alone or in combinations; growth regulator herbicide used for control of weeds in cereals. *Bayer AG. See* 2,4-D MCPA, dichlorprop, MCPP.

947   Hytrol
Aminotriazole-2,4-D + diuron + simazine.
Aminotriazole, 2,4-D, diuron and simazine; used for total weed control in non crop areas. *Farmers Crop Chemicals Ltd. See* aminotriazole, 2,4-D, diuron, simazine.

948   Hytrol
2,4-D-Diuron + amitrole + simazine.
2,4-D, diuron, amitrole, simazine; total weedkiller

with more than one seasons resistance; for use on garden paths and other noncrop areas. *Agrichem (International) Ltd. See* 2,4-D, diuron, amitrole, simazine.

949   Keytrol
A total weedkiller containing aminotriazole, atrazine and 2,4-D in a wettable formulation; provides broad spectrum control of grassy and broad-leaf weed species, including deep-rooted perennials. *Burts & Harvey. See* aminotriazole; atrazine and 2,4-D.

950   Legumex Extra
A solution concentrate containing 27 g benazolin, 237 g 2,4-DB and 42.3 g MCPA per liter; a post-emergence herbicide. *Schering Agrochemicals Ltd.*

951   Longlife Plus
Mixture of 2,4-D and dicamba; used to control weeds in turf. *ICI Agrochemicals Professional Products. See* 2,4-D, dicamba.

952   Malezafin LV-4
Emulsifiable concentrate of butoxyethanol ester of 2,4-D acid; low volatile, broadleaf herbicide for corn crops and pasture land. *Invequimica & CIA SCA. Se* 2,4-D.

953   Malezafin 57 LV
Emulsifiable concentrate of a mixture of butoxyethanol esters of 2,4-D and dichlorprop; low volatile, broad leaf herbicide, brush killer for pasture land. *Invequimica & CIA SCA. See* butoxyethanol esters of 2,4-D and dichlorprop.

954   Mascot Selective
2,4-D + mecoprop.
Soluble concentrate containing 60 g 2,4-D and 200 g mecoprop per liter; used to control weeds in grassland. *Rigby Taylor Ltd. See* mecoprop.

955   MSS 2,4-DB + MCPA
2,4-DB [94-75-7] and MCPA [94-74-6]; translocated herbicide applied to cereals and undersown clovers. *Mirfield Sales Services Ltd. See* 2,4-DB, MCPA.

956   New Formulation SBK Brushwood Killer
Liquid concentrate containing 2,4-D, mecoprop and dicamba; selective herbicide for use on coarse and woody weeds. *Vitax Ltd. See* 2,4-D, mecoprop, dicamba.

957   Palormone
Aqueous solution containing 50% w/v 2,4-D as the amine salt; for use as an agricultural herbicide. *Universal Crop Protection Ltd. See* 2,4-D.

958   Polymone
94-75-7; 120-36-5
Aqueous solution containing 10% w/v 2,4-D and 40% w/v Dichlorprop; for use as an agricultural herbicide. *Universal Crop Protection Ltd.*

959   Primatol AD 85WP
Aminotriazole + atrazine + 2,4-D; used for total weed control in non crop areas. *Ciba-Geigy Agrochemicals.*

960   Select-Trol
Soluble concentrate containing 6.6% w/w 2,4-D and 250 g mecoprop; used to control weeds in grassland. *Chemsearch (UK) Ltd.*

961   Setter 33
A solution concentrate containing 50 g benazolin, 237 g 2,4-DB and 43 g MCPA per liter; a post-emergence herbicide. *DowElanco Ltd.* See benazolin, 2,4-DB, 43 g MCPA.

962   Snapper CDA
An emulsifiable concentrate containing 95 g aminotriazole, 190 g atrazine and 99 g 2,4-D per liter; used for total weed control in non crop areas. *ICI Agrochemicals Professional Products.* See aminotriazole, atrazine, 2,4-D.

963   Super Verdone
Soluble concentrate containing 72 g 2,4-D, 12 g dicamba and 48 g ioxynil per liter; used to control weeds in turf. *ICI Chem & Polymers Ltd.* See dicamba, ioxynil.

964   Supertox
2,4-D + mecoprop; for control weeds in grassland. *Rhône-Poulenc/Agri.* See 2,4-D, mecoprop.

965   Sydex
Soluble concentrate containing 125 g 2,4-D and 250 g mecoprop per liter; used to control weeds in grassland. *Synchemicals Ltd.*

966   Terramix CDA
An emulsifiable concentrate containing 99 g aminotriazole, 190 g atrazine and 95 g 2,4-D per liter; used for total weed control in non crop areas. *Denoon CDS; Powaspray (CDA) Ltd.*

967   Terranox
Emulsifiable concentrate containing 105g aminotriazole, 207g atrazine and 100g 2,4-D per liter; for total weed control in non crop areas. *Agri-Technics Ltd.*

968   Topshot
Bentazone + cyanazine + 2,4-DB. Herbicide. *Shell UK.*

969   Torapron
An emulsifiable concentrate containing 95 g aminotriazole, 190 g atrazine and 99 g 2,4-D per liter; used for total weed control in non crop areas. *BP Oil Ltd.* See aminotriazole, atrazine, 2,4-D.

970   Trik
Aminotriazole + 2,4-D + diuron; used for total weed control in non crop areas. *Smyth-Morris Chemicals Ltd.* See aminotriazole, 2,4-D, diuron.

971   Verdone 2
Contains mecoprop and 2,4-D; selective lawn weedkiller. *ICI Garden Products.*

972   Verdone CDA
94-75-7; 7058-19-0   2802,5666   202-361-1
$C_8H_6Cl_2O_3$; $C_{10}H_{11}ClO_3$
2,4-D + Mecoprop.
[2,4-D]: (2,4-dichlorophenoxy)acetic acid; Hedonal; Trinoxol; [Mecoprop]: (±)-2-(4-chloro-2-methylphenoxy)propanoic acid; (±)-2-[(4-chloro-o-tolyl)oxy]-propionic acid; mechlorprop; MCPP; CMPP; RD 4593; Astix CMPP; IsoCornox; Compitox; Compitox Plus; Proponex-Plus. Emulsifiable concentrate containing 6.7% 2,4-D, and 13.3% mecoprop; used to control weeds in grassland. [2,4-D]: mp = 138°; $bp_{0.4}$ = 160°; [Mecoprop]: mp = 95-96°; $[\alpha]_D^{25}$ = +19 (alcohol). *ICI Agrochemicals Professional Products.*

973   Weed and Brushkiller
Emulsifiable concentrate containing 144 g 2,4-D, 32 g dicamba and 144 g mecoprop per liter; an herbicide to control perennial and woody weeds. *Synchemicals Ltd.* See 2,4-D, dicamba, mecoprop.

974   Weedkill
Aminotriazole + 2,4-D + diuron + simazine; used for total weed control in non crop areas. *Dermaglen Ltd.* See aminotriazole, 2,4-D, diuron, simazin.

975   Zennapron
Mixture of 2,4-D and mecoprop. Used to control weeds in grassland. *BP Oil Ltd.* See 2,4-D, mecoprop.

Herbicides containing Aminotriazole

976   Atlazin
Atrazine + aminotriazole.
A liquid formulation containing 250g atrazine [1912-24-9] and 218g aminotriazole [61-82-5] per liter as a suspension concentrate; used for total weed control on industrial sites, paths, kerbs and channels, drives and hard tennis courts, hardstanding and storage areas. *Chipman Ltd.*

977 Atraflow Plus

A liquid formulation containing 264g atrazine [1912-24-9] and 214g aminotriazole [61-82-5] per liter as a suspension concentrate; for used where total weed control is required including industrial sites, paths, curbs, and channels, drives and hard tennis courts, hardstanding and storage areas. *Burts & Harvey. See* atrazine, aminotriazole.

978 Atraflow Plus

A liquid formulation containing 270g atrazine [1912-24-9] and 160g aminotriazole [61-82-5] per liter as a suspension concentrate; for use where total weed control is required including industrial sites, paths, curbs, and channels, drives and hard tennis courts, hard standing and storage areas. *Rhône-Poulenc Environmental Prods. Ltd. See* atrazine, aminotriazole.

979 Boroflow S/ATA

A suspension concentrate containing 160 g aminotriazole and 270 g simazine per liter; used for total weed control in non crop areas and fruit orchards. *ABM Chemicals Ltd. See* aminotriazole, simazine.

980 Bullseye CDA

Amitrole, atrazine and diuron; a liquid mixture of herbicides for weed control. *ICI Agrochemicals. See* amitrole, atrazine, diuron.

981 CDA Simflow Plus

A suspension concentrate containing 100 g aminotriazole [61-82-5] and 300 g simazine [122-34-9] per liter; used for total weed control in non crop areas and fruit orchards. *Rhône-Poulenc Environmental Prods. Ltd. See* aminotriazole, simazine.

982 CDA Viper

Amitrole-atrazine-diuron; a liquid mixture of herbicides for weed control. *CDA Chemicals Ltd. See* amitrole, atrazine, diuron.

983 Clearway

Aminotriazole + simazine.
A suspension concentrate containing 100 g aminotriazole [61-82-5] and 300 g simazine [122-34-9] per liter; used for total weed control in non crop areas and fruit orchards. *Rhône-Poulenc Environmental Prods. Ltd. See* aminotriazole, simazine.

984 Duranox

Amitrole + atrazine + diuron.
Amitrole, atrazine and diuron; a liquid mixture of herbicides for weed control. *Agri-Technics Ltd.*

985 Herbazin Plus

Simazine + aminotriazole.
A wettable powder containing simazine and aminotriazole; a quick acting herbicide for control of existing weeds with long term persistence. *Fisons plc, Horticultural Div.*

986 Herbazin Plus SC

Amitrole + simazine.
A suspension concentrate containing 180 g aminotriazole [61-82-5] and 300 g simazine [122-34-9] per liter; used for total weed control in non crop areas and fruit orchards. *Fisons plc. See* amitrole, simazine.

987 Hytrol

Aminotriazole + 2,4-D + diuron + simazine.
Aminotriazole, 2,4-D, diuron and simazine; used for total weed control in non crop areas. *Farmers Crop Chemicals Ltd. See* aminotriazole, 2,4-D, diuron, simazine.

988 Hytrol

2,4-D-Diuron + amitrole + simazine.
2,4-D, diuron, amitrole, simazine; total weedkiller with more than one seasons resistance; for use on garden paths and other noncrop areas. *Agrichem (International) Ltd. See* 2,4-D, diuron, amitrole, simazine.

989 Kagolin 5.8FG

Amitrole + atrazine + diuron; a granular mixture of herbicides for weed control. *Ciba-Geigy Agrochemicals. See* amitrole; atrazine; diuron.

990 Keytrol

A total weedkiller containing aminotriazole, atrazine and 2,4-D in a wettable formulation; provides broad spectrum control of grassy and broad-leaf weed species, including deep-rooted perennials. *Burts & Harvey. See* aminotriazole; atrazine and 2,4-D.

991 Mascot Highway

Aminotriazole [61-82-5] and simazine [122-34-9]; used for total weed control in non crop areas. *Rigby Taylor Ltd. See* aminotriazole, simazine.

992 Meganox Plus

An emulsifiable concentrate containing 100 g aminotriazole, 250 g atrazine and 100 g MCPA per liter; used for total weed control in non crop areas. *Agri-Technics Ltd. See* aminotriazole, atrazine, MCPA.

993   MSS Simazine/Aminotriazole 43FL

A suspension concentrate containing 155 g [122-34-9] per liter; used for total weed control in non crop areas and fruit orchards. *Mirfield Sales Services Ltd. See aminotriazole, simazine.*

994   Pathclear

Contains aminotriazole, diquat, paraquat and simazine; long-acting weedkiller for paths, drives, and patios. *ICI Garden Products.*

995   Primatol AD 85WP

Aminotriazole + atrazine + 2,4-D; used for total weed control in non crop areas. *Ciba-Geigy Agrochemicals.*

996   Sarapron

An emulsifiable concentrate containing 98.8 g aminotriazole, 197.5 g atrazine and 20 g dicamba per liter; used for total weed control in non crop areas. *BP Oil Ltd. See aminotriazole, atrazine, dicamba.*

997   Serramix CDA

An emulsifiable concentrate containing 100 g aminotriazole, 200 g atrazine and 100 g MCPA per liter; used for total weed control in non crop areas. *Denoon CDS; Powaspray (CDA) Ltd. See aminotriazole, atrazine, MCPA.*

998   Simazol

Active ingredients: azolan (aminotriazole) plus simanex (simazine); multipurpose herbicidal mixture which eradicates a wide spectrum of established weeds, while preventing further weed germination for extended periods. *Agan Chemical Manufacturers Ltd. See simazine, aminotriazole.*

999   Simflow Plus

A suspension concentrate containing 100 g aminotriazole [61-82-5] and 300 g simazine [122-34-9] per liter; used for total weed control in non crop areas and fruit orchards. *Rhone-Poulenc Environmental Prods. Ltd. See aminotriazole, simazine.*

1000   Snapper CDA

An emulsifiable concentrate containing 95 g aminotriazole, 190 g atrazine and 99 g 2,4-D per liter; used for total weed control in non crop areas. *ICI Agrochemicals Professional Products.*

1001   Synchemicals Total Weed Killer

A suspension concentrate containing 53 g aminotriazole [61-82-5] and 110 g simazine [122-34-9] per liter; used for total weed control in non crop areas and fruit orchards. *Synchemicals Ltd. See aminotriazole, simazine.*

1002   Syntox Total Weed Killer

A suspension concentrate containing 53 g aminotriazole [61-82-5] and 100 g simazine [122-34-9] per liter; used for total weed control in non crop areas and fruit orchards. *Syntex Manufacturing Ltd. See aminotriazole, simazine.*

1003   Terramix CDA

An emulsifiable concentrate containing 99 g aminotriazole, 190 g atrazine and 95 g 2,4-D per liter; used for total weed control in non crop areas. *Denoon CDS; Powaspray (CDA) Ltd. See aminotriazole, atrazine, 2,4-D.*

1004   Terranox

Emulsifiable concentrate containing 105g aminotriazole, 207g atrazine and 100g 2,4-D per liter; For total weed control in non crop areas. *Agri-Technics Ltd. See aminotriazole, atrazine, 2,4-D.*

1005   Torapron

An emulsifiable concentrate containing 95 g aminotriazole, 190 g atrazine and 99 g 2,4-D per liter; used for total weed control in non crop areas. *BP Oil Ltd. See aminotriazole, atrazine, 2,4-D.*

1006   Trik

Aminotriazole + 2,4-D + diuron; used for total weed control in non crop areas. *Smyth-Morris Chemicals Ltd. See aminotriazole, 2,4-D, diuron.*

1007   Ustinex®

Products with combinations of different herbicidal compounds (aminotriazole, diuron, methabenzthiazuron, phenoxies); for weed control on paths, open spaces, parks and sports grounds. *Bayer AG. See aminotriazole, diuron, methabenzthiazuron.*

1008   Weedkill

Aminotriazole + 2,4-D + diuron + simazine; used for total weed control in non crop areas. *Dermaglen Ltd.*

Herbicides containing Atrazine

1009   Arsenal XL

A soluble concentrate containing 300g atrazine and 12.5 g imazapyr per liter; used for total weed control in non crop areas. *Chipman Ltd.*

1010   Arsenal® XL

Soluble concentrate containing 300 g atrazine and 12.5g imazapyr per liter; for total weed control in noncrop areas. *Cyanamid of Great Britain Ltd.*

1011   Atlacide Extra

Atrazine-sodium chlorate; used for total weed control in non crop areas. *Chipman Ltd. See atrazine, sodium chlorate.*

1012   Atlas Lignum Granules

Atrazine-dalapon; granular soil acting herbicide for use in forestry plantations. *Atlas Interlates Ltd. See* atrazine, dalapon.

1013   Atlazin

Atrazine + aminotriazole.

A liquid formulation containing 250g atrazine [1912-24-9] and 218g aminotriazole [61-82-5] per liter as a suspension concentrate; used for total weed control on industrial sites, paths, kerbs and channels, drives and hard tennis courts, hardstanding and storage areas. *Chipman Ltd.*

1014   Atraflow Plus

A liquid formulation containing 264g atrazine [1912-24-9] and 214g aminotriazole [61-82-5] per liter as a suspension concentrate; for used where total weed control is required including industrial sites, paths, curbs, and channels, drives and hard tennis courts, hardstanding and storage areas. *Burts & Harvey. See* atrazine, aminotriazole.

1015   Atraflow Plus

A liquid formulation containing 270g atrazine [1912-24-9] and 160g aminotriazole [61-82-5] per liter as a suspension concentrate; for use where total weed control is required including industrial sites, paths, curbs, and channels, drives and hard tennis courts, hard standing and storage areas. *Rhône-Poulenc Environmental Prods. Ltd. See* atrazine, aminotriazole.

1016   Atramet Combi

Active ingredients: atranex plus ametrex, ready formulated mixture of atrazine plus ametryne for use as a selective pre-and post-emergence herbicide. *Agan Chemical Manufacturers Ltd. See* atrazine, ametryne.

1017   Borocil Extra

Atrazine + boromacil + diuron.

Atrazine, bromacil and diuron; used for total weed control in non crop areas. *ABM Chemicals Ltd.*

1018   Bullseye CDA

Amitrole, atrazine and diuron; a liquid mixture of herbicides for weed control. *ICI Agrochemicals.*

1019   CDA Viper

Amitrole-atrazine-diuron; a liquid mixture of herbicides for weed control. *CDA Chemicals Ltd. See* amitrole, atrazine, diuron.

1020   Duranox

Amitrole + atrazine + diuron.

Amitrole, atrazine and diuron; a liquid mixture of herbicides for weed control. *Agri-Technics Ltd. See* amitrole, atrazine, diuron.

1021   Gardoprim A 500FW

Atrazine + terbuthylazine.

A suspension concentrate containing 100 g atrazine and 400 g terbuthylazine per liter; used for total weed control in forestry plantations. *Ciba-Geigy Agrochemicals. See* atrazine, terbuthylazine.

1022   Holtox

Atrazine + cyanazine.

A suspension concentrate containing 250 g atrazine and 250 g cyanazine per liter; a residual herbicide. *Shell UK. See* atrazine, cyanazine.

1023   Kagolin 5.8FG

Amitrole + atrazine + diuron; a granular mixture of herbicides for weed control. *Ciba-Geigy Agrochemicals. See* amitrole; atrazine; diuron.

1024   Keytrol

A total weedkiller containing aminotriazole, atrazine and 2,4-D in a wettable formulation; provides broad spectrum control of grassy and broad-leaf weed species, including deep-rooted perennials. Burts & Harvey. *See* aminotriazole; atrazine and 2,4-D.

1025   Meganox Plus

An emulsifiable concentrate containing 100 g aminotriazole, 250 g atrazine and 100 g MCPA per liter; used for total weed control in non crop areas. *Agri-Technics Ltd. See* aminotriazole, atrazine, MCPA.

1026   Moderator

A soluble concentrate containing 300 g atrazine and 12.5 g imazapyr per liter; used for total weed control in non crop areas. *Chipman Ltd. See* atrazine, imazapyr.

1027   Primatol AD 85WP

Aminotriazole + atrazine + 2,4-D; for total weed control in non crop areas. *Ciba-Geigy Agrochemicals. See* aminotriazole, atrazine, 2,4-D.

1028   Sarapron

An emulsifiable concentrate containing 98.8 g aminotriazole, 197.5 g atrazine and 20 g dicamba per liter; used for total weed control in non crop areas. *BP Oil Ltd. See* aminotriazole, atrazine, dicamba.

1029   Serramix CDA

An emulsifiable concentrate containing 100 g aminotriazole, 200 g atrazine and 100 g MCPA per liter; used for total weed control in non crop areas. *Denoon CDS; Powaspray (CDA) Ltd. See* aminotriazole, atrazine, MCPA.

1030 Snapper CDA

An emulsifiable concentrate containing 95 g aminotriazole, 190 g atrazine and 99 g 2,4-D per liter; used for total weed control in non crop areas. *ICI Agrochemicals Professional Products.* See aminotriazole, atrazine, 2,4-D.

1031 Terramix CDA

An emulsifiable concentrate containing 99 g aminotriazole, 190 g atrazine and 95 g 2,4-D per liter; used for total weed control in non crop areas. *Denoon CDS; Powaspray (CDA) Ltd.* See aminotriazole, atrazine, 2,4-D.

1032 Terranox

An emulsifiable concentrate containing 105g aminotriazole, 207g atrazine and 100g 2,4-D per liter; used for total weed control in non crop areas. *Agri-Technics Ltd.* See aminotriazole, atrazine, 2,4-D.

1033 Torapron

An emulsifiable concentrate containing 95 g aminotriazole, 190 g atrazine and 99 g 2,4-D per liter; used for total weed control in non crop areas. *BP Oil Ltd.* See aminotriazole, atrazine, 2,4-D.

Herbicides containing Benazolin

1034 Asset

An emulsifiable concentrate containing 50 g benazolin, 125g bromoxynil and 62.5 g ioxynil per liter; a post-emergence herbicide for cereal crops and grass. *Schering Agrochemicals Ltd.* See benazolin, bromoxynil and ioxynil.

1035 Jaguar

Benazolin + bromoxynil + ioxynil + mecoprop. An emulsifiable concentrate containing 22.2 g benazolin, 55.6 g bromoxynil, 27.8 g ioxynil and 413 g mecoprop per liter; a post-emergence herbicide for cereal crops and grass. *Schering Agrochemicals Ltd.* See benazolin, bromoxynil, ioxynil, mecoprop.

1036 Legumex Extra

A solution concentrate containing 27 g benazolin, 237 g 2,4-DB and 42.3 g MCPA per liter; a post-emergence herbicide. *Schering Agrochemicals Ltd.*

1037 Setter 33

A solution concentrate containing 50 g benazolin, 237 g 2,4-DB and 43 g MCPA per liter; a post-emergence herbicide. *DowElanco Ltd.* See benazolin, 2,4-DB, 43 g MCPA.

Herbicides containing Bentazon

1038 Acumen

Bentazone + MCPA + MCPB.

Post-emergence contact and translocated herbicide for undersown cereals. *BASF plc.* See bentazone, MCPA, MCPB.

1039 Galaxy®

Bentazon + acifluorfen.
For post-emergence control of annual broadleaf weeds in soybeans and peanuts. *BASF AG.* See bentazon.

1040 Herbatox®

Bentazon + isoproturon + dichlorprop.
Bentazon, isoproturon, dichlorprop; for postemergence control of grasses and broadleaf weeds in winter cereals and spring wheat. *BASF AG.* See bentazon, isoproturon, dichlorprop.

1041 Pulsar

A solution concentrate containing 200 g bentazone and 200 g MCPA per liter; a post-emergence herbicide. *BASF plc.* See bentazone, MCPA.

1042 Topshot

Bentazone + cyanazine + 2,4-DB; an herbicide. *Shell UK.* See bentazone, cyanazine, 2,4-DB.

1043 Triagran®

Bentazon, dichorprop, MCPA; post-emergence herbicide for control of broad-leaved weeds in winter and spring cereals. *BASF AG.* See bentazon, dichorprop, MCPA.

Herbicides containing Bifenox

1044 Alibi

Herbicide containing bifenox and linuron. *ICI Chem & Polymers Ltd.* See linuron.

1045 Dicurane Duo 495FW

Bifenox + chlorotoluron.
A liquid formulation containing 106 g bifenox and 389 g chlorotoluron per liter as a suspension concentrate; a residual herbicide for the control of weeds in winter wheat. *Ciba-Geigy Agrochemicals.* See bifenox, chlorotoluron.

1046 Foxstar

Bifenox + isoproturon + mecoprop.
A liquid formulation containing 107 g bifenox, 286 g isoproturon and 143 g mecoprop per liter as a suspension concentrate; a post-emergence herbicide for the control of weeds in winter cereals. *Rhône-Poulenc Crop Protection Ltd.* See bifenox, isoproturon, mecoprop.

1047 Invicta Duo 495FW

Bifenox + isoproturon.
A liquid formulation containing 160 g bifenox and 400 g isoproturon per liter as a suspension

concentrate; a residual herbicide for the control of weeds in winter cereals. *Farmers Crop Chemicals Ltd. See* bifenox, isoproturon.

1048   Invicta Duo 495FW
Bifenox + isoproturon.
A liquid formulation containing 160 g bifenox and 400 g isoproturon per liter as a suspension concentrate; a residual herbicide for the control of weeds in winter cereals. *Farm Protection Ltd. See* bifenox, isoproturon.

Herbicides containing Bromacil

1049   Borocil Extra
Atrazine + boromacil + diuron.
Atrazine, bromacil and diuron; used for total weed control in non crop areas. *ABM Chemicals Ltd.*

1050   Borocil K
Bromacil + diuron.
Bromacil and diuron; used for total weed control in non crop areas. *ABM Chemicals Ltd.*

1051   Fenocil
Bromacil + pentachlorophenol.
Bromacil and pentachlorophenol; used for total weed control in noncrop areas. *Chemsearch (UK) Ltd. See* bromacil, pentachlorophenol.

1052   Hydon
Bromacil + picloram.
Bromacil and picloram; used for total weed control in non crop areas. *Chipman Ltd. See* bromacil, picloram.

1053   Krovar
A wettable powder containing 40% w/w bromacil and 40% w/w diuron; used for total weed control in noncrop areas. *Selectokil Ltd. See* bromacil and diuron.

1054   Krovar®
A wettable powder containing 40% w/w bromacil and 40% w/w diuron; used for total weed control in noncrop areas. *DuPont UK. See* bromacil and diuron.

Herbicides containing Bromoxynil

1055   Advance
Bromoxynil, fluroxypyr and ioxynil post-emergence contact herbicide for cereals. Broad-spectrum herbicide. *(Sold in UK for Dow Elanco). ICI Chem & Polymers Ltd.; ICI Agrochemicals.*

1056   Asset
An emulsifiable concentrate containing 50 g benazolin, 125g bromoxynil and 62.5 g ioxynil per liter; a post-emergence herbicide for cereal crops and grass. *Schering Agrochemicals Ltd. See* benazolin, bromoxynil and ioxynil.

1057   Boxolon
Herbicide containing clopyralid, bromoxynil and mecoprop. *ICI Chem & Polymers Ltd.*

1058   Briotril
Bromoxynil octanoate and ioxynil octanoate; herbicide used for selective post-emergence weed control. *Agan Chemical Manufacturers Ltd.*

1059   Briotril Plus
Bromoxynil octanoate + ioxynil octanoate.
An emulsifiable concentrate containing 200 g bromoxynil and 200 g ioxynil per liter; a post-emergence contact herbicide for cereal crops. *Pan Britannica Industries Ltd. See* bromoxynil, ioxynil.

1060   Bromotrill
Bromoxynil octanoate + 2,6-dibromo-4-cyanophenyl octanoate.
Active ingredients: bromoxynil octanoate and 2,6-dibromo-4-cyanophenyl octanoate; selective postemergence control of a wide range of annual broadleaf weeds in winter and spring cereals and in com. *Agan Chemical manufacturers Ltd. See* bromoxynil octanoate, 2,6-dibromo-4-cyanophenyl octanoate.

1061   Chafer Certrol-E
Bromoxynil, dichlorprop and ioxynil; herbicide mixture for weed control in spring cereals. *DuPont UK. See* bromoxynil, dichlorprop, ioxynil.

1062   Deloxil
Bromoxynil + oxynil.
An emulsifiable concentrate containing 190 g bromoxynil and 190 g ioxynil per liter; a post-emergence contact herbicide for cereal crops. *Hoechst UK. See* bromoxynil, ioxynil.

1063   Hobane
Bromoxynil [1689-84-5] + ioxynil [16849-83-4].
Bromoxynil and ioxynil; herbicide. *ICI Chem & Polymers Ltd. See* bromoxynil, ioxynil.

1064   Hobane
Bromoxynil [1689-84-5] + ioxynil [16849-83-4].
An emulsifiable concentrate containing 240 g bromoxynil [1689-84-5] and 160 g ioxynil [1689-33-4] per liter; a post-emergence contact herbicide for cereal crops. *Farm Protection Ltd. See* bromoxynil, ioxynil.

1065   Jaguar
Benazolin + bromoxynil + ioxynil + mecoprop.
An emulsifiable concentrate containing 22.2 g benazolin, 55.6 g bromoxynil, 27.8 g ioxynil and 413 g mecoprop per liter; a post-emergence

herbicide for cereal crops and grass. *Schering Agrochemicals Ltd. See* benazolin, bromoxynil, ioxynil, mecoprop.

1066   Nortron Leyclene

Bromoxynil, ethofumesate, ioxynil; herbicide mixture for new grass lays. *Schering Agrochemicals Ltd. See* bromoxynil, ethofumesate, ioxynil.

1067   Novacorn

An emulsifiable concentrate containing 240 g bromoxynil and 160 g ioxynil per liter; a post-emergence contact herbicide for cereal crops. *Farmers Crop Chemicals Ltd. See* bromoxynil, ioxynil.

1068   Oxytril CM

An emulsifiable concentrate containing 200 g bromoxynil and 200 g ioxynil per liter; a post-emergence contact herbicide for cereal crops. *Rhône-Poulenc Crop Protection Ltd. See* bromoxynil, ioxynil.

1069   Sickle

Bromoxynil + fluroypyr; post-emergence contact herbicide for cereals. *DowElanco Ltd. See* bromoxynil, bromoxynil + fluroypyr.

1070   Stellox 380EC

An emulsifiable concentrate containing 190 g bromoxynil and 190 g ioxynil per liter; a post-emergence contact herbicide for cereal crops. *Ciba-Geigy Agrochemicals. See* bromoxynil, ioxynil.

1071   Swipe 560 EC

An emulsifiable concentrate containing 56 g bromoxynil, 56 g loxynil and 448 g mecoprop per liter; a post-emergence contact herbicide for cereal crops. *Ciba-Geigy Agrochemicals. See* bromoxynil, ioxynil, mecoprop.

1072   Terset

Bromoxynil + ioxynil + isoproturon + mecoprop; a contact herbicide for use in cereal crops. *Rhône-Poulenc Crop Protection Ltd.*

1073   Tetroxone

Selective weedkiller containing bromoxynil, dichlorprop, ioxynil and MCPA as potassium salts. *ICI Chem & Polymers Ltd. See* bromoxynil, dichlorprop, ioxynil, MCPA.

1074   Vindex

Bromoxynil + clopyralid; herbicide mixture for weed control in cereals. *Quadrangle Agrochemicals. See* bromoxynil, clopyralid.

1075   Vulcan

Herbicide containing clopyralid and bromoxynil. *ICI Chem & Polymers Ltd. See* clopyralid, bromoxynil.

Herbicides containing Cetrimide

1076   Croptex Pewter

cetrimide + chlorpropham.
A suspension concentrate containing 80 g cetrimide (hexadecyltrimethylammonium bromide) and 80 g chlorpropham per liter; soil acting herbicide for lettuce. *Hortichem Ltd. See* cetrimide, chlorpropham.

1077   Hebron Pabracr

Cetrimide + chlorpropham.
A suspension concentrate containing 80 g cetrimide and 80 g chlorpropham per liter. soil acting herbicide for lettuce. *Atlas Interlates Ltd. See* cetrimide, chlorpropham.

Herbicides containing Chloridazon

1078   Advizor

Chloridazon and lenacil; pre-emergence herbicide for use in sugar beet. *ICI Chem & Polymers Ltd. See* chloridazon, lenacil.

1079   Ashlade CP

Suspension concentrate containing 86 g chloridazon and 400 g propachlor per liter; a residual herbicide for beet crops. *Ashlade Formulations Ltd. See* chloridazon, propachlor.

1080   Atlas Electrum

Suspension concentrate containing 200 g chloridazon, 30 g chlorpropham, 20 g fenuron and 120 g propham per liter; a residual herbicide for beet crops. *Atlas Interlates Ltd. See* chloridazon, chlorpropham, fenuron, propham.

1081   Barrier

Suspension concentrate containing 300 g chloridazon, 30 g fenuron and 170g propham per liter; a residual herbicide for beet crops. *Truchem Ltd.*

1082   Magnum® F

Mixture of chloridazon and ethofumesate; selective systemic herbicide. *BASF AG; BASF plc. See* chloridazon, ethofumesate.

1083   Pyradur®

Chloridazon + metolachlor; pre-emergence herbicide for control of grasses and broad-leaved weeds in sugar and fodder beet crops. *BASF AG. See* chloridazon, metolachlor.

1084   Spectron

Suspension concentrate containing 211 g chloridazon [1698-60-8] and 200 g ethofumesate [26225-79-6] per liter; a residual herbicide for beet crops. *Schering Agrochemicals Ltd. See* chloridazon, ethofumesate.

Herbicides containing Chlorpropham

1085   Atlas Brown

Emulsifiable concentrate containing 150 g chlorpropham and 100 g pentanochlor per liter, a residual herbicide. *Atlas Interlates Ltd. See* chlorpropham, pentanochlor.

1086   Atlas Electrum

Suspension concentrate containing 200 g chloridazon, 30 g chlorpropham, 20 g fenuron and 120 g propham per liter; a residual herbicide for beet crops. *Atlas Interlates Ltd. See* chloridazon, chlorpropham, fenuron, propham.

1087   Atlas Red

Suspension concentrate containing 200 g chlorpropham, cresylic acid and 50 g fenuron per liter; an herbicide for use on ornamentals and vegetables. *Atlas Interlates Ltd. See* chlorpropham, cresylic acid, fenuron.

1088   Croptex Chrome

chlorpropham + fenuron.
Suspension concentrate containing 80 g chlorpropham and 15 g fenuron per liter, an herbicide for use on ornamentals and vegetables. *Hortichem Ltd.*

1089   Croptex Pewter

cetrimide + chlorpropham.
A suspension concentrate containing 80 g cetrimide (hexadecyltrimethylammonium bromide) and 80 g chlorpropham per liter; soil acting herbicide for lettuce. *Hortichem Ltd.*

1090   Hebron Pabracr

Cetrimide + chlorpropham.
A suspension concentrate containing 80 g cetrimide and 80 g chlorpropham per liter. soil acting herbicide for lettuce. *Atlas Interlates Ltd.*

1091   MSS Sugar Beet Herbicide

A suspension concentrate containing 37.5 g chlorpropham, 25 g fenuron and 150 g propham per liter; an herbicide for use on beet crops. *Mirfield Sales Services Ltd.*

1092   Profalon

Emulsifiable concentrate containing 200 g chlorpropham and 100 g linuron per liter; used to control weeds in bulb crops. *Hoechst UK.*

1093   Residuren Extra

Chlorpropham [101-21-3] and diuron [330-54-1]; an herbicide for treatment of grass in bulbs, peas, and beans. *Farm Protection Ltd. See* chlorpropham, diuron.

1094   Truchem Quintex

Emulsifiable concentrate containing 30 g chlorpropham, 30 g fenuron and 130 g propham per liter; an herbicide for beet crops. *MTM Agrochemicals Ltd. See* chlorpropham, fenuron, propham.

Herbicides containing Clopyralid

1095   Boxolon

Herbicide containing clopyralid, bromoxynil and mecoprop. *ICI Chem & Polymers Ltd.*

1096   Coupler SC

clopyralid + cyanazine.
Suspension concentrate containing 60 g clopyralid and 350 g cyanazine per liter; a contact herbicide. *Shell UK. See* clopyralid, cyanazine.

1097   Grazon 90

Clopyralid + triclopyr.
Emulsifiable concentrate containing 60 g clopyralid and 240 g triclopyr per liter; used for treatment of perennial weeds in grassland. *Dow Elanco Ltd. See* clopyralid, triclopyr.

1098   Hotspur

Clopyralid + fluroxypyr + ioxynil.
Mixture of clopyralid, fluroxypyr and ioxynil; used to control broad leaf weeds in cereals. *Farm Protection Ltd. See* clopyralid, fluroxypyr, ioxynil.

1099   Lontrel Plus

Soluble concentrate containing 15 g clopyralid, 420 g dichlorprop and 175 g MCPA per liter; a translocated herbicide for use on cereals. *(Sold in UK for DowElanco). ICI Chem & Polymers Ltd.*

1100   Matrikerb

Mixture of clopyralid and propyzamide; a soil and leaf herbicide for winter oilseed rape. *Pan Britannica Industries Ltd.*

1101   Seloxone

Soluble concentrate containing 15 g clopyralid and 510 g mecoprop per liter; a translocated herbicide for cereals and grassland. *ICI Chem & Polymers Ltd.*

1102   Vindex

Bromoxynil + clopyralid; herbicide mixture for weed control in cereals. *Quadrangle Agrochemicals.*

1103   Vulcan
Herbicide containing clopyralid and bromoxynil.
ICI Chem & Polymers Ltd.

Herbicides containing Cyanazine

1104   Cleaval
Cyanazine + mecoprop.
Suspension concentrate containing 60 g cyanazine
and 400 g mecoprop per liter; a contact herbicide.
Shell UK. See cyanazine, mecoprop.

1105   Coupler SC
Clopyralid + cyanazine.
Suspension concentrate containing 60 g clopyralid
and 350 g cyanazine per liter; a contact herbicide.
Shell UK. See clopyralid, cyanazine.

1106   Holtox
Atrazine + cyanazine.
A suspension concentrate containing 250 g
atrazine and 250 g cyanazine per liter; a residual
herbicide. *Shell UK. See* atrazine, cyanazine.

1107   Quiver
Suspension concentrate containing 200 g
cyanazine and 350 g isoproturon per liter; a
residual herbicide for winter cereals. *Shell UK. See*
cyanazine, isoproturon.

1108   Spitfire
Mixture of cyanazine and fluroxypyr; a post-
emergence herbicide. *DowElanco Ltd.*

1109   Topshot
Bentazone + cyanazine + 2,4-DB; an herbicide.
Shell UK. See bentazone, cyanazine, 2,4-DB.

1110   Laser
Emulsifiable concentrate containing 200 g
cycloxydim per liter; used to control weeds in
grass. *BASF plc.*

Herbicides containing Dalapon

1111   Atlas Lignum Granules
Atrazine-dalapon; granular soil acting herbicide
for use in forestry plantations. *Atlas Interlates Ltd.*
See atrazine, dalapon.

1112   Destral
2,4-D + Dalapon + diuron.
Mixture of 2,4-D, dalapon and diuron; used for
total weed control in non crop areas. *ABM
Chemicals Ltd. See* 2,4-D, dalapon, diuron.

1113   Volunteered
Mixture of dalapon and di-1-p-menthene; a grass
weed herbicide. *Mandops (UK) Ltd.*

Herbicides containing Dicamba

1114   Banlene Plus
Soluble concentrate of 18g dicamba, 252g MCPA
and 84g mecoprop per liter; used for weed control
in cereals and grassland. *Schering Agrochemicals
Ltd. See* dicamba, MCPA, mecoprop.

1115   BH Dockmaster
Soluble concentrate of 125g dicamba and 125g
maleic hydrazide per liter; used to control docks
in road verges and noncrop grass. *Rhône-Poulenc
Environmental Prods. Ltd. See* dicamba, maleic
hydrazide.

1116   Broadshot
Emulsifiable concentrate containing 200 g 2,4-D,
85 g dicamba and 65 g triclopyr per liter; an
herbicide to control perennial and woody weeds.
Shell UK. See 2,4-D, dicamba, triclopyr.

**1117   Campbell's Field Marshal, Grassland
Herbicide**
Mixture of dicamba, MCPA and mecoprop; used
for weed control in cereals and grassland. *MTM
AgroChemicals Ltd. See* dicamba, MCPA,
mecoprop.

1118   Di-Farmon
Dicamba + mecoprop.
Soluble concentrate of 21 g dicamba and 319g
mecoprop per liter; used for weed control in
cereals and grassland. *ICI Chem & Polymers Ltd;
Farm Protection Ltd. See* dicamba, mecoprop.

1119   Dock-Ban
Dicamba + MCPA + mecoprop.
Soluble concentrate of 19.5 g dicamba, 245 g
MCPA and 86.5 g mecoprop per liter; used for
weed control in cereals and grassland.
Quadrangle Agrochemicals. See dicamba, MCPA,
mecoprop.

1120   Docklene
Dicamba + MCPA + mecoprop.
Soluble concentrate of 336 g mecoprop, 84 g
dicamba and 84 g MCPA per liter; used for weed
control in cereals and grassland. *Schering
Agrochemicals Ltd.*

1121   Endox
      3090
Dicamba + mecoprop.
Soluble concentrate of 112 g dicamba [1918-00-
9] and 265 g mecoprop 7085-19-0] per liter;
herbicide used for weed control in cereals and
grassland. *Farmers Crop Chemicals Ltd.*

1122   Estermone
2,4-D + dicamba.
Mixture of 2,4-D and dicamba; used to control weeds in turf. *Synchemicals Ltd. See* 2,4-D, dicamba.

1123   Farmon Condox
Dicamba + mecoprop.
Soluble concentrate of 112 g dicamba and 265 g mecoprop per liter; used for weed control in cereals and grassland. *Farm Protection Ltd. See* dicamba, mecoprop.

1124   Fettel
Dicamba + mecoprop + triclopyr.
Emulsifiable concentrate of 78g dicamba, 130g mecoprop, and 72g triclopyr per liter; used for weed control in cereals and grassland. *Farm Protection Ltd. See* dicamba, mecoprop, triclopyr.

1125   Green Up
2,4-D + dicamba.
Mixture of 2,4-D and dicamba; used to control weeds in turf. *Synchemicals Ltd. See* 2,4-D, dicamba.

1126   Green Up Lawn Feed and Weed
Liquid concentrate containing NPK 14.5:3:3 and 2,4-D and dicamba; combined feed and weed for turf. *Vitax Ltd.*

**1127   Green Up Lawn Feedn Weed Plus
            Moss Killer**
Dichlorophen + mecoprop + dichlorprop + dicamba + benazolin.
Liquid concentrate containing NK 11:4 plus dichlorophen, mecoprop, dichlorprop, dicamba and benazolin; combined feed, weed and mosskiller. *Vitax Ltd. See* dichlorophen, mecoprop, dichlorprop, dicamba, benazolin.

1128   Green Up Weedfree Lawn Weedkiller
2,4-D + dicamba.
Emulsifiable concentrate containing 2,4-D [1702-17-6] and dicamba [1918-00-9]; selective herbicide for use on turf. *Vitax Ltd.*

**1129   Green Up Weedfree Spot Weedkiller
            for Lawns**
2,4-D + dicamba.
Aerosol containing 2,4-D [1702-17-6] and dicamba [1918-00-9]; selective spot weedkiller for lawns. *Vitax Ltd.*

1130   Headland Relay
Dicamba + MCPA + mecoprop.
Mixture of dicamba, MCPA and mecoprop; used for weed control in cereals and grassland. *WBC Technology Ltd. See* dicamba, MCPA, mecoprop.

1131   Herrisol
Dicamba + MCPA + mecoprop.
A liquid containing 35.4% w/w dicamba (Banvel D), MCPA and mecoprop; used to control common chickweed, cleavers, knotgrass, redshank, scentless mayweed, corn spurrey and a wide range of other broadleaved weeds in cereals, grass crops and orchards. *Bayer plc.*

1132   Holdfast D
Dicamba + paclobutrazol.
Soluble concentrate of 25 g dicamba and 250 g paclobutrazol per liter; used for weed control in cereals and grassland. *ICI Agrochemicals. See* dicamba, paclobutrazol.

1133   Hyban
Dicamba + mecoprop.
Soluble concentrate of 18.7 g dicamba and 300 g mecoprop per liter; used for weed control in cereals and grassland. *Agrichem (International) Ltd. See* dicamba, mecoprop.

1134   Hygrass
Dicamba + mecoprop.
Soluble concentrate of 18.7 g dicamba and 300 g mecoprop per liter; used for weed control in cereals and grassland. *Agrichem (International) Ltd. See* dicamba, mecoprop.

1135   Hyprone
Dicamba + MCPA + mecoprop.
Soluble concentrate of 18.7 g dicamba, 100 g MCPA and 194 g mecoprop per liter, used for weed control in cereals and grassland. *Agrichem (International) Ltd.*

1136   Hysward
Dicamba + MCPA + mecoprop.
Mixture of dicamba, MCPA and mecoprop; used for weed control in cereals and grassland. *Agrichem (International) Ltd.*

1137   Longlife Plus
Mixture of 2,4-D and dicamba; used to control weeds in turf. *ICI Agrochemicals Professional Products.*

1138   Mascot Super Selective
Soluble concentrate of 15 g dicamba, 100 g MCPA and 200 g mecoprop per liter; used for weed control in cereals and grassland. *Rigby Taylor Ltd. See* mecoprop, dicamba, MCPA.

1139   Mephetol Extra
Soluble concentrate of 20.8 g dicamba, 333 g dichloprop and 200 g MCPA per liter; used for weed control in cereals and newly sown grass. *BritAg Industries Ltd.*

1141 MSS Mircam

Soluble concentrate of 18.7 g dicamba and 300 g mecoprop per liter; used for weed control in cereals and grassland. *Mirfield Sales Services Ltd.*

1142 MSS Mircam Plus

Soluble concentrate of 19.5 g dicamba, 245 g MCPA and 86.5 g mecoprop per liter; used for weed control in cereals and grassland. *Mirfield Sales Services Ltd. See* dicamba, MCPA, mecoprop.

1143 New Formulation SBK Brushwood Killer

Liquid concentrate containing 2,4-D, mecoprop and dicamba; selective herbicide for use on coarse and woody weeds. *Vitax Ltd. See* 2,4-D, mecoprop, dicamba.

1144 Paddox

Herbicide containing the sodium/potassium salts of MCPA, mecoprop, and dicamba. *ICI Chem & Polymers Ltd. See* mecoprop, dicamba.

1145 Paddox

Soluble concentrate of 18 g dicamba, 237 g MCPA, and 80 g mecoprop per liter; used for weed control in cereals and grassland. *Farm Protection Ltd. See* mecoprop, dicamba.

1146 Pasturol Plus

Soluble concentrate of 25 g dicamba, 200 g MCPA, and 400 g mecoprop per liter; broad-spectrum herbicide used for weed control in cereals and grassland. *Farmers Crop Chemicals Ltd. See* mecoprop.

1147 Quadban

Soluble concentrate of 19.5 g dicamba, 245 g MCPA and 86.5 g mecoprop per liter; used for weed control in cereals and grassland. *Quadrangle Agrochemicals. See* dicamba, MCPA, mecoprop.

1148 Sarapron

An emulsifiable concentrate containing 98.8 g aminotriazole, 197.5 g atrazine and 20 g dicamba per liter; used for total weed control in non crop areas. *BP Oil Ltd. See* aminotriazole, atrazine, dicamba.

1149 Selectrol

Soluble concentrate of 18 g dicamba, 252 g MCPA and 84 g mecoprop per liter; used for weed control in cereals and grassland. *R P Adams Ltd. See* dicamba, MCPA, mecoprop.

1150 Springcorn Extra

Soluble concentrate of 18 g dicamba, 360 g MCPA and 160 g mecoprop per liter; used for weed control in cereals and grassland. *Farmers Crop Chemicals Ltd. See* dicamba, MCPA, mecoprop.

1151 Super Verdone

Soluble concentrate containing 72 g 2,4-D, 12 g dicamba and 48 g ioxynil per liter; used to control weeds in turf. *ICI Chem & Polymers Ltd. See* dicamba, ioxynil.

1152 Tetralex Plus

Soluble concentrate of 18 g dicamba, 252 g MCPA and 84 g mecoprop per liter; used for weed control in cereals and grassland. *Shell UK. See* dicamba, MCPA, mecoprop.

1153 Tribute

Mixture of dicamba, MCPA and mecoprop; used for weed control in cereals and grassland. *Chipman Ltd. See* dicamba, MCPA, mecoprop.

1154 Tritox

An aqueous concentrate containing MCPA, mecoprop, and dicamba; a selective herbicide for the control of broad-leaved weeds in turf. *Fisons plc, Horticultural Div. See* MCPA, mecoprop, dicamba.

1155 Weed and Brushkiller

Emulsifiable concentrate containing 144 g 2,4-D, 32 g dicamba and 144 g mecoprop per liter; an herbicide to control perennial and woody weeds. *Synchemicals Ltd. See* 2,4-D, dicamba, mecoprop.

Herbicides containing Dichlorophen

1156 Green Up Lawn Feedn Weed Plus Moss Killer

Dichlorophen + mecoprop + dichlorprop + dicamba + benazolin.
Liquid concentrate containing NK 11:4 plus dichlorophen, mecoprop, dichlorprop, dicamba and benazolin; combined feed, weed and mosskiller. *Vitax Ltd. See* dichlorophen, mecoprop, dichlorprop, dicamba, benazolin.

1157 SHL Lawn Sand Plus

Dichlorophen + ferrous sulfate; moss killer/fertilizer mixture for turf. *Sinclair Horticulture & Leisure Ltd. See* dichlorophen, ferrous sulfate.

Herbicides containing Dichlorprop

1158 Chafer Certrol-E

Bromoxynil, dichlorprop and ioxynil; herbicide mixture for weed control in spring cereals. *DuPont UK. See* bromoxynil, dichlorprop, ioxynil.

1159   Farmon 2,4-DP.CPA

Dichlorprop + MCPA.
Soluble concentrate of 358 g dichlorprop and 177 g MCPA per liter; used for control of weeds in cereals and grass. *Farm Protection Ltd.*

1160   Green Up Lawn Feedn Weed Plus Moss Killer

Dichlorophen + mecoprop + dichlorprop + dicamba + benazolin.
Liquid concentrate containing NK 11:4 plus dichlorophen, mecoprop, dichlorprop, dicamba and benazolin; combined feed, weed and mosskiller. *Vitax Ltd. See* dichlorophen, mecoprop, dichlorprop, dicamba, benazolin.

1161   Hedonal®

2,4-D MCPA + dichlorprop + MCPP.
Range of herbicides containing 2,4-D MCPA, dichlorprop, MCPP either alone or in combinations; growth regulator herbicide used for control of weeds in cereals. *Bayer AG. See* 2,4-D MCPA, dichlorprop, MCPP.

1162   Hemoxone

Dichlorprop + MCPA.
Soluble concentrate of 392 g dichlorprop and 210g MCPA per liter; used for control of weeds in cereals and grass. *ICI Chem & Polymers Ltd. See* dichlorprop, MCPA.

1163   Herbatox®

Bentazon + isoproturon + dichlorprop.
Bentazon, isoproturon, dichlorprop; for postemergence control of grasses and broadleaf weeds in winter cereals and spring wheat. *BASF AG. See* bentazon, isoproturon, dichlorprop.

1164   Lontrel Plus

Soluble concentrate containing 15 g clopyralid, 420 g dichlorprop and 175 g MCPA per liter; a translocated herbicide for use on cereals. (Sold in UK for DowElanco). *ICI Chem & Polymers Ltd. See* clopyralid, dichloprop, MCPA.

1165   Malezafin 57 LV

Emulsifiable concentrate of a mixture of butoxyethanol esters of 2,4-D and dichlorprop; low volatile, broad leaf herbicide, brush killer for pasture land. *Invequimica & CIA SCA. See* butoxyethanol esters of 2,4-D and dichlorprop.

1166   MSS 2,4-DP + MCPA

Mixture of dichlorprop and MCPA; used for weed control in cereals and turf. *Mirfield Sales Services Ltd. See* dichlorprop, MCPA.

1167   MSS CMPP/DP

Soluble concentrate of 520 g dichlorprop and 130 g mecoprop per liter; used for control of weeds in barley, wheat and oats. *Mirfield Sales Services Ltd. See* dichlorprop, mecoprop.

1168   Polymone

94-75-7; 120-36-5
Aqueous solution containing 10% w/v 2,4-D and 40% w/v Dichlorprop; for use as an agricultural herbicide. *Universal Crop Protection Ltd. See* 2,4-D, dichlorprop.

1169   Seritox

Mixture of dichlorprop and MCPA; used for weed control in cereals and turf. *Rhone-Poulenc Crop Protection Ltd. See* dichlorprop and MCPA.

1170   Tetroxone

Selective weedkiller containing bromoxynil, dichlorprop, ioxynil and MCPA as potassium salts. *ICI Chem & Polymers Ltd.*

1171   Triagran®

Bentazon, dichorprop, MCPA; post-emergence herbicide for control of broad-leaved weeds in winter and spring cereals. *BASF AG. See* bentazon, dichorprop, MCPA.

Herbicides containing Diflufenican

1172   Ardent

Suspension concentrate containing 40 g diflufenican and 400 g trifluralin per liter; used for control of weeds in winter cereals. *Embetec Crop Protection Ltd. See* diflufenican, trifluralin.

1173   Javelin

Diflufenican + isoproturon.
Suspension concentrate containing 62.5 g diflufenican and 500 g isoproturon per liter; used for control of weeds in winter cereals. *Rhone-Poulenc Crop Protection Ltd.*

1174   Panther

Suspension concentrate containing 50 g diflufenican and 500 g isoproturon per liter; used for control of weeds in winter cereals. *Rhone-Poulenc Crop Protection Ltd.*

Herbicides containing Diquat

1175   Dukatalon

Diquat + paraquat.
Mixture of diquat and paraquat; herbicide. *Makhteshim Chemical Works Ltd.*

1176   Farmon PDQ

Diquat + paraquat.
Soluble concentrate of 80 g diquat and 120 g paraquat per liter; used for weed control in field crops. *Farm Protection Ltd. See* diquat, paraquat.

1177   Parable
Soluble concentrate of 100 g diquat and 100 g paraquat per liter; used for weed control in field crops. *ICI Agrochemicals. See* diquat, paraquat.

1178   Pathclear
Contains aminotriazole, diquat, paraquat and simazine; long-acting weedkiller for paths, drives, and patios. *ICI Garden Products. See* aminotriazole, diquat, paraquat , simazine.

1179   Soltair
Mixture of diquat, paraquat, and simazine; total herbicide. *ICI Chem & Polymers Ltd. See* diquat, paraquat, simazine.

1180   Weedol
Weedkiller containing paraquat [1910-42-5] and diquat [85-00-7] for gardeners. *ICI Garden Products. See* paraquat, diquat.

Herbicides containing Diuron

1181   Borocil Extra
Atrazine + boromacil + diuron.
Atrazine, bromacil and diuron; used for total weed control in non crop areas. *ABM Chemicals Ltd. See* atrazine, bromacil and diuron.

1182   Borocil K
Bromacil + diuron.
Bromacil and diuron; used for total weed control in non crop areas. *ABM Chemicals Ltd. See* bromacil, diuron.

1183   Bullseye CDA
Amitrole, atrazine and diuron; a liquid mixture of herbicides for weed control. ICI Agrochemicals. *See* amitrole, atrazine, diuron.

1184   CDA Viper
Amitrole-atrazine-diuron; a liquid mixture of herbicides for weed control. *CDA Chemicals Ltd. See* amitrole, atrazine, diuron.

1185   Destral
2,4-D + dalapon + diuron.
Mixture of 2,4-D, dalapon and diuron; used for total weed control in non crop areas. *ABM Chemicals Ltd. See* 2,4-D, dalapon, diuron.

1186   Dexuron
Diuron + paraquat.
Suspension concentrate containing 300 g diuron and 100 g paraquat per liter; used to control weeds around trees and shrubs. *Chipman Ltd.*

1187   Diurol
Azolan + diurex.
Active   ingredients: azolan   plus   diurex;
multlpurpose herbicidal mixture which eradicates a wide spectrum of established weeds while preventing further weed germination for extended periods. *Agan Chemical Manufacturers Ltd. See* azolan, diurex.

1188   Duranox
Amitrole + atrazine + diuron.
Amitrole, atrazine and diuron; a liquid mixture of herbicides for weed control. *Agri-Technics Ltd. See* amitrole, atrazine, diuron.

1189   Gramixel
Paraquat + diuron.
Herbicide containing paraquat [1910-42-5] and diuron [330-54-1]. *ICI Chem & Polymers Ltd. See* paraquat, diuron.

1190   Gramuron
Paraquat + diuron.
Herbicide containing paraquat [1910-42-5] and diuron [330-54-1]. *ICI Chem & Polymers Ltd. See* paraquat, diuron.

1191   Hytrol
Aminotriazole + 2,4-D + diuron + simazine.
Aminotriazole, 2,4-D, diuron and simazine; used for total weed control in non crop areas. *Farmers Crop Chemicals Ltd. See* aminotriazole, 2,4-D, diuron, simazine.

1192   Hytrol
2,4-D + diuron + amitrole + simazine.
2,4-D, diuron, amitrole, simazine; total weedkiller with more than one seasons resistance; for use on garden paths and other noncrop areas. *Agrichem (International) Ltd. See* 2,4-D, diuron, amitrole, simazine.

1193   Kagolin 5.8FG
Amitrole + atrazine + diuron; a granular mixture of herbicides for weed control. *Ciba-Geigy Agrochemicals. See* amitrole, atrazine, diuron.

1194   Krovar
A wettable powder containing 40% w/w bromacil and 40% w/w diuron; used for total weed control in noncrop areas. *Selectokil Ltd. See* bromacil, diuron.

1195   Krovar®
A wettable powder containing 40% w/w bromacil and 40% w/w diuron; used for total weed control in noncrop areas. *DuPont UK. See* bromacil, diuron.

1196   Paracol
Herbicide containing diuron and paraquat. ICI Chem & Polymers Ltd. *See* diuron, paraquat.

1197   Residuren Extra

Chlorpropham [101-21-3] and diuron [330-54-1]; an herbicide for treatment of grass in bulbs, peas, and beans. *Farm Protection Ltd. See* chlorpropham, diuron.

1198   Simfix

Granular mixture of diuron and simazine; used for control of weeds in woody crops and noncrop areas. *Rhône-Poulenc Environmental Prods. Ltd. See* diuron, simazine.

1199   Totacol

Herbicide based on diuron and paraquatermary. *ICI Chem & Polymers Ltd. See* diuron, paraquat.

1200   Trik

Aminotriazole + 2,4-D + diuron; used for total weed control in non crop areas. *Smyth-Morris Chemicals Ltd. See* aminotriazole, 2,4-D, diuron.

1201   Uradex

Active ingredients; diurex plus uragan; selective pre-emergence herbicide mixture. *Agan Chemical Manufacturers Ltd. See* diurex, uragan.

1202   Ustinex®

Products with combinations of different herbicidal compounds (aminotriazole, diuron, methabenz-thiazuron, phenoxies); herbicides used for control of weeds on paths, open spaces, parks and sports grounds. *Bayer AG. See* aminotriazole, diuron, methabenzthiazuron.

1203   Weedkill

Aminotriazole + 2,4-D + diuron + simazine; used for total weed control in non crop areas. *Dermaglen Ltd.*

Herbicides containing Ethofumesate

1204   Magnum® F

Mixture of chloridazon and ethofumesate; selective systemic herbicide. *BASF AG; BASF plc. See* chloridazon, ethofumesate.

1205   Nortron Leyclene

Bromoxynil, ethofumesate, ioxynil; herbicide mixture for new grass lays. *Schering Agrochemicals Ltd. See* bromoxynil, ethofumesate, ioxynil.

1206   Spectron

Suspension concentrate containing 211 g chloridazon [1698-60-8] and 200 g ethofumesate [26225-79-6] per liter; a residual herbicide for beet crops. *Schering Agrochemicals Ltd. See* chloridazon, ethofumesate.

1207   Malezafin 55 Plus

Emulsifiable concentrate of ethyl ester of 2,4-D acid and butoxyethanol ester of dichlorprop; broadleaf herbicide for pasture land. *Invequimica & CIA SCA. See* ethyl ester of 2,4-D acid and butoxyethanol ester of dichlorprop.

Herbicides containing Fenuron

1208   Atlas Electrum

Suspension concentrate containing 200 g chloridazon, 30 g chlorpropham, 20 g fenuron and 120 g propham per liter; a residual herbicide for beet crops. *Atlas Interlates Ltd. See* chloridazon, chlorpropham, fenuron, propham.

1209   Atlas Red

Suspension concentrate containing 200 g chlorpropham, cresylic acid and 50 g fenuron per liter; an herbicide for use on ornamentals and vegetables. *Atlas Interlates Ltd. See* chlorpropham, cresylic acid, fenuron.

1210   Barrier

Suspension concentrate containing 300 g chloridazon, 30 g fenuron and 170g propham per liter; a residual herbicide for beet crops. *Truchem Ltd.*

1211   Croptex Chrome

chlorpropham + fenuron.
Suspension concentrate containing 80 g chlorpropham and 15 g fenuron per liter, an herbicide for use on ornamentals and vegetables. *Hortichem Ltd. See* chlorpropham, fenuron.

1212   MSS Sugar Beet Herbicide

A suspension concentrate containing 37.5 g chlorpropham, 25 g fenuron and 150 g propham per liter; an herbicide for use on beet crops. *Mirfield Sales Services Ltd.*

1213   Truchem Quintex

Emulsifiable concentrate containing 30 g chlorpropham, 30 g fenuron and 130 g propham per liter; an herbicide for beet crops. *MTM Agrochemicals Ltd. See* chlorpropham, fenuron, propham.

Herbicides containing Ferrous Sulfate

1214   Ashlade D-Moss

Mixture of chloroxuron and ferrous sulfate; a lawn sand herbicide to control mosses in turf. *Ashlade Formulations Ltd. See* chloroxuron, ferrous sulfate.

1215 Green Up Feed and Weed Plus Mosskiller

Dry powder containing NPK 8:4:4 plus 2,4-D, mecoprop and ferrous sulfate; combined fertilizer, weed and mosskiller. *Vitax Ltd.*

1216 SHL Lawn Sand Plus

Dichlorophen + ferrous sulfate; moss killer/fertilizer mixture for turf. *Sinclair Horticulture & Leisure Ltd.*

Herbicides containing Fluroxypyr

1217 Advance

Bromoxynil, fluroxypyr and ioxynil post-emergence contact herbicide for cereals. Broad-spectrum herbicide. *(Sold in UK for Dow Elanco). ICI Chem & Polymers Ltd.; ICI Agrochemicals. See* bromoxynil, fluroxypyr, ioxynil.

1218 Hotspur

Clopyralid + fluroxypyr + ioxynil.
Mixture of clopyralid, fluroxypyr and ioxynil; used to control broad leaf weeds in cereals. *Farm Protection Ltd. See* clopyralid, fluroxypyr, ioxynil.

1219 Spitfire

Mixture of cyanazine and fluroxypyr; a post-emergence herbicide. *DowElanco Ltd. See* cyanazine, fluroxypyr.

1220 Sickle

Bromoxynil + fluroypyr; post-emergence contact herbicide for cereals. *DowElanco Ltd. See* bromoxynil, bromoxynil + fluroypyr.

Herbicides containing Glyphosate

1221 Bronco

Active ingredients are 2.6 lb of 2-chloro-2',6'-diethyl-N-(methoxymethyl) acetanilide and 1.4 lb of the isopropylamine salt of glyphosate; herbicide for no-till farming. *Monsanto Co.*

1222 Mascot Ultrasonic

Glyphosate + simazine.
Glyphosate + simazine; a translocated and residual herbicide for the control of grasses and weeds. *Rigby Taylor Ltd. See* glyphosate, simazine.

Herbicides containing Imazapyr

1223 Arsenal XL

A soluble concentrate containing 300g atrazine and 12.5 g imazapyr per liter; used for total weed control in non crop areas. *Chipman Ltd. See* atrazine, imazapyr.

1224 Arsenal® XL

Soluble concentrate containing 300 g atrazine and 12.5g imazapyr per liter; for total weed control in noncrop areas. *Cyanamid of Great Britain Ltd. See* atrazine, imazapyr.

1225 Moderator

A soluble concentrate containing 300 g atrazine and 12.5 g imazapyr per liter; used for total weed control in non crop areas. *Chipman Ltd. See* atrazine, imazapyr.

Herbicides containing Ioxynil

1226 Advance

Bromoxynil, fluroxypyr and ioxynil post-emergence contact herbicide for cereals. Broad-spectrum herbicide. *(Sold in UK for Dow Elanco). ICI Chem & Polymers Ltd.; ICI Agrochemicals. See* bromoxynil, fluroxypyr, ioxynil.

1227 Asset

An emulsifiable concentrate containing 50 g benazolin, 125g bromoxynil and 62.5 g ioxynil per liter; a post-emergence herbicide for cereal crops and grass. *Schering Agrochemicals Ltd. See* benazolin, bromoxynil and ioxynil.

1228 Briotril

Bromoxynil octanoate and ioxynil octanoate; herbicide used for selective post-emergence weed control. *Agan Chemical Manufacturers Ltd. See* bromoxynil, ioxynil.

1229 Briotril Plus

Bromoxynil octanoate + ioxynil octanoate.
An emulsifiable concentrate containing 200 g bromoxynil and 200 g ioxynil per liter; a post-emergence contact herbicide for cereal crops. *Pan Britannica Industries Ltd. See* bromoxynil, ioxynil.

1230 Chafer Certrol-E

Bromoxynil, dichlorprop and ioxynil; herbicide mixture for weed control in spring cereals. *DuPont UK. See* bromoxynil, dichlorprop, ioxynil.

1231 Deloxil

bromoxynil + oxynil.
An emulsifiable concentrate containing 190 g bromoxynil and 190 g ioxynil per liter; a post-emergence contact herbicide for cereal crops. *Hoechst UK. See* bromoxynil, oxynil.

1232 Harrier

Mecoprop-3,6-dichloropicolinic acid + ioxynil.
Weedkiller containing mecoprop, 3,6-dichloropicolinic acid and ioxynil as potassium salts. *ICI Chem & Polymers Ltd. See* mecoprop, 3,6-dichloropicolinic acid, ioxynil.

1233   Hobane
Bromoxynil [1689-84-5] + ioxynil [16849-83-4].
Bromoxynil and ioxynil; herbicide. *ICI Chem & Polymers Ltd. See* bromoxynil, ioxynil.

1234   Hobane
Bromoxynil [1689-84-5] + ioxynil [16849-83-4]. An emulsifiable concentrate containing 240 g bromoxynil [1689-84-5] and 160 g ioxynil [1689-33-4] per liter; a post-emergence contact herbicide for cereal crops. *Farm Protection Ltd.*

1235   Hotspur
Clopyralid + fluroxypyr + ioxynil.
Mixture of clopyralid, fluroxypyr and ioxynil; used to control broad leaf weeds in cereals. *Farm Protection Ltd. See* clopyralid, fluroxypyr, ioxynil.

1236   Iotox
Ioxynil + mecoprop.
Soluble concentrate containing 72 g ioxynil and 214 g mecoprop per liter; used for weed control in turf. *Rhône-Poulenc Environmental Prods. Ltd.*

1237   Jaguar
Benazolin + bromoxynil + ioxynil + mecoprop.
An emulsifiable concentrate containing 22.2 g benazolin, 55.6 g bromoxynil, 27.8 g ioxynil and 413 g mecoprop per liter; a post-emergence herbicide for cereal crops and grass. *Schering Agrochemicals Ltd.*

1238   Musketeer
Suspension concentrate containing 50 g ioxynil, 250 g isoproturon and 180 g mecoprop per liter; used for control of weeds in wheat and barley. *Hoechst UK. See* oxynil, isoproturon, mecoprop.

1239   Nortron Leyclene
Bromoxynil, ethofumesate, ioxynil; herbicide mixture for new grass lays. *Schering Agrochemicals Ltd. See* bromoxynil, ethofumesate, ioxynil.

1240   Novacorn
An emulsifiable concentrate containing 240 g bromoxynil and 160 g ioxynil per liter; a post-emergence contact herbicide for cereal crops. *Farmers Crop Chemicals Ltd. See* bromoxynil, ioxynil.

1241   Oxytril CM
An emulsifiable concentrate containing 200 g bromoxynil and 200 g ioxynil per liter; a post-emergence contact herbicide for cereal crops. *Rhône-Poulenc Crop Protection Ltd. See* bromoxynil, ioxynil.

1242   Post-Kite
Suspension concentrate containing 50 g ioxynil,

250 g isoproturon and 180 g mecoprop per liter; used for control of weeds in wheat and barley. *Schering Agro-chemicals Ltd. See* ioxynil, isoproturon, mecoprop.

1243   Stellox 380EC
An emulsifiable concentrate containing 190 g bromoxynil and 190 g ioxynil per liter; a post-emergence contact herbicide for cereal crops. *Ciba-Geigy Agrochemicals. See* bromoxynil, ioxynil.

1244   Super Verdone
Soluble concentrate containing 72 g 2,4-D, 12 g dicamba and 48 g ioxynil per liter; used to control weeds in turf. *ICI Chem & Polymers Ltd. See* dicamba, ioxynil.

1245   Synox
Soluble concentrate containing 75 g ioxynil and 225 g mecoprop per liter; used for weed control in turf. *Synchemicals Ltd. See* ioxynil, mecoprop.

1246   Terset
Bromoxynil + ioxynil + isoproturon + mecoprop; a contact herbicide for use in cereal crops. *Rhône-Poulenc Crop Protection Ltd. See* bromoxynil, ioxynil, isoproturon, mecoprop.

1247   Tetroxone
Selective weedkiller containing bromoxynil, dichlorprop, ioxynil and MCPA as potassium salts. *ICI Chem & Polymers Ltd. See* bromoxynil, dichlorprop, ioxynil, MCPA.

Herbicides containing Isoproturon

1248   Autumn Kite
Emulsifiable concentrate of 300 g isoproturon [34123-59-6] and 200 g trifluralin [1582-09-8] per liter; used for control of annual grasses in winter wheat and barley. *Schering Agrochemicals Ltd.*

1249   Encore
Isoproturon + pendimethalin.
Suspension concentrate containing 125 g isoproturon and 250 g pendimethalin per liter; used for annual weed control in winter wheat, rye and barley. *Cyanamid of Great Britain Ltd.*

1250   Fanfare
Isoproturon + isoxaben.
Suspension concentrate containing 450 g isoproturon and 19 g isoxaben per liter; used for annual weed control in cereals. *Ciba-Geigy Agrochemicals. See* isoproturon, isoxaben.

1251   Foxstar
Bifenox + isoproturon + mecoprop.
A liquid formulation containing 107 g bifenox,

286 g isoproturon and 143 g mecoprop per liter as a suspension concentrate; a post-emergence herbicide for the control of weeds in winter cereals. *Rhône-Poulenc Crop Protection Ltd.*

1252 Herbatox®

Bentazon + isoproturon + dichlorprop.
Bentazon, isoproturon, dichlorprop; for postemergence control of grasses and broadleaf weeds in winter cereals and spring wheat. *BASF AG.*

1253 Invicta Duo 495FW

Bifenox + isoproturon.
A liquid formulation containing 160 g bifenox and 400 g isoproturon per liter as a suspension concentrate; a residual herbicide for the control of weeds in winter cereals. *Farmers Crop Chemicals Ltd. See* bifenox, isoproturon.

1254 Invicta Duo 495FW

Bifenox + isoproturon.
A liquid formulation containing 160 g bifenox and 400 g isoproturon per liter as a suspension concentrate; a residual herbicide for the control of weeds in winter cereals. *Farm Protection Ltd. See* bifenox, isoproturon.

1255 Ipso

Isoproturon + isoxaben.
Suspension concentrate containing 450 g isoproturon and 19 g isoxaben per liter; used for annual weed control in cereals. *DowElanco Ltd.*

1256 Javelin

Diflufenican + isoproturon.
Suspension concentrate containing 62.5 g diflufenican and 500 g isoproturon per liter; used for control of weeds in winter cereals. *Rhône-Poulenc Crop Protection Ltd.*

1257 Musketeer

Suspension concentrate containing 50 g ioxynil, 250 g isoproturon and 180 g mecoprop per liter; used for control of weeds in wheat and barley. *Hoechst UK. See* oxynil, isoproturon, mecoprop.

1258 Oracle®

Isoproturon and metsulfuron-methyl; used for annual weed control in wheat and barley. *DuPont UK. See* isoproturon, metsulfuron-methyl.

1259 Panther

Suspension concentrate containing 50 g diflufenican and 500 g isoproturon per liter; used for control of weeds in winter cereals. *Rhône-Poulenc Crop Protection Ltd. See* diflufenican, isoproturon.

1260 Pinnacle

Mixture of imazamethabenz-methyl and isoproturon; used for control of weed grasses in winter cereals. *Cyanamid of Great Britain Ltd. See* imazamethabenz-methyl, isoproturon.

1261 Post-Kite

Suspension concentrate containing 50 g ioxynil, 250 g isoproturon and 180 g mecoprop per liter; used for control of weeds in wheat and barley. *Schering Agro-chemicals Ltd. See* ioxynil, isoproturon, mecoprop.

1262 Quiver

Suspension concentrate containing 200 g cyanazine and 350 g isoproturon per liter; a residual herbicide for winter cereals. *Shell UK. See* cyanazine, isoproturon.

1263 Terset

Bromoxynil + ioxynil + isoproturon + mecoprop; a contact herbicide for use in cereal crops. *Rhône-Poulenc Crop Protection Ltd. See* bromoxynil, ioxynil, isoproturon, mecoprop.

1264 Trump

Suspension concentrate containing 236 g isoproturon and 236 g pendimethalin per liter; used for annual weed control in winter wheat, rye, and barley. *Cyanamid of Great Britain Ltd. See* isoproturon, pendimethalin.

Herbicides containing Isoxaben

1265 Fanfare

Isoproturon + isoxaben.
Suspension concentrate containing 450 g isoproturon and 19 g isoxaben per liter; used for annual weed control in cereals. *Ciba-Geigy Agrochemicals. See* isoproturon, isoxaben.

1266 Glytex

Isoxaben + methabenzthiazuron.
Mixture of isoxaben and methabenzthiazuron; soil-acting herbicide for winter cereals. *Bayer plc. See* isoxaben, methabenzthiazuron.

1267 Ipso

Isoproturon + isoxaben.
Suspension concentrate containing 450 g isoproturon and 19 g isoxaben per liter; used for annual weed control in cereals. *DowElanco Ltd. See* isoproturon, isoxaben.

Herbicides containing Lenacil

1268   Advizor
Chloridazon and lenacil; pre-emergence herbicide for use in sugar beet. *ICI Chem & Polymers Ltd.* *See* chloridazon, lenacil.

1269   DUK-880
Lenacil + phenmedipham.
Lenacil and phenmedipham; used for control of annual dicotyledons in sugar beet. *DuPont UK. See* lenacil, phenmedipham.

Herbicides containing Linuron

1270   Alibi
Herbicide containing bifenox and linuron. *ICI Chem & Polymers Ltd.*

1271   Alistell
Linuron + 2,4-DB + MCPA.
Herbicide containing linuron, 2,4-DB and MCPA. *ICI Chem & Polymers Ltd.*

1272   Bronox
Wettable powder containing linuron and tritrazine; used for control of annual dicotyledons and annual grasses in potatoes and nursery stock. *Schering Agrochemicals Ltd. See* linuron and tritrazine.

1273   Campbell's Trifluron
Mixture of linuron [330-55-2] and trifluralin [1582-09-8]; herbicide for winter cereals. *MTM AgroChemicals Ltd. See* linuron, trifluralin.

1274   Chandor
Emulsifiable concentrate containing 120 g linuron and 240 g trifluralin per liter; herbicide for winter cereals. *DowElanco Ltd. See* linuron, trifluralin.

1275   Clovacorn Extra
An emulsifiable concentrate containing 220 g 2,4-DB, 30 g linuron and 30 g MCPA per liter; used to control weeds in undersown cereals and seedling grassland. *Farmers Crop Chemicals Ltd. See* 2,4-DB, linuron, MCPA.

1276   Flint
Linuron + trifluralin.
Emulsifiable concentrate containing 120 g linuron and 240 g trifluralin per liter; herbicide for winter cereals. *Ashlade Formulations Ltd.*

1277   Janus
Linuron + trifluralin.
Liquid mixture of linuron [330-55-2] and trifluralin [1982-09-8]; herbicide for winter cereals. *Atlas Interlates Ltd. See* linuron, trifluralin.

1278   Marksman
Mixture of linuron and trifluralin; herbicide for winter cereals. *Farmers Crop Chemicals Ltd. See* linuron and trifluralin.

1279   Onslaught
Suspension concentrate containing 160 g linuron [330-55-2] and 320 g trifluralin [1582-09-8] per liter; herbicide for winter cereals. *Quadrangle Agrochemicals. See* linuron, trifluralin.

1280   Pre-Empt
Emulsifiable concentrate containing 46 g linuron, 54 g trietazine and 200 g trifluralin per liter; herbicide for winter cereals. *Schering Agrochemicals Ltd. See* linuron, trietazine, trifluralin.

1281   Premaline
Herbicide based upon linuron. *May & Baker Ltd.*

1282   Profalon
Emulsifiable concentrate containing 200 g chlorpropham and 100 g linuron per liter; used to control weeds in bulb crops. *Hoechst UK. See* chlorpropham, linuron.

1283   Solo
Emulsifiable concentrate containing 1250 g linuron [330-55-2] and 240 g trifluralin [1582-09-8] per liter; herbicide for winter cereals. *MTM Agrochemicals Ltd. See* linuron, trifluralin.

1284   Swipe 560 EC
An emulsifiable concentrate containing 56 g bromoxynil, 56 g loxynil and 448 g mecoprop per liter; a post-emergence contact herbicide for cereal crops. *Ciba-Geigy Agrochemicals. See* bromoxynil, loxynil, mecoprop.

1285   Tempo
Herbicide containing linuron and terbutryn. *ICI Chem & Polymers Ltd. See* linuron, terbutryn.

1286   Tempo
Suspension concentrate containing 150 g linuron and 150 g terbutryn per liter; used for control of annual dicotyledons in potatoes. *Farm Protection Ltd. See* linuron, terbutryn.

1287   Trifarmon
Suspension concentrate containing 160 g linuron [330-55-2] and 320 g trifluralin [1582-09-8] per liter; herbicide for winter cereals. Farm Protection Ltd. *See* linuron, trifluralin.

1288   Tri-Farmon
Herbicide containing linuron [330-55-2] and trifluralin [1582-09-8]. *ICI Chem & Polymers Ltd.*

Herbicides containing MCPA

1289 Acumen

Bentazone + MCPA + MCPB.
Post-emergence contact and translocated herbicide for undersown cereals. *BASF plc. See* Betazone, MCPA, MCPB.

1290 Agrichem DB Plus

2,4-DB + MCPA; translocatable herbicide for cereal crops. *Agrichem (International) Ltd. See* 2,4-DB, MCPA.

1291 Alistell

Linuron + 2,4-DB + MCPA.
Herbicide containing linuron, 2,4-DB and MCPA. *ICI Chem & Polymers Ltd.*

1292 Banlene Plus

Soluble concentrate of 18g dicamba, 252g MCPA and 84g mecoprop per liter; used for weed control in cereals and grassland. *Schering Agrochemicals Ltd. See* dicamba, MCPA, mecoprop.

1293 Campbell's Field Marshal, Grassland Herbicide

Mixture of dicamba, MCPA and mecoprop; used for weed control in cereals and grassland. *MTM AgroChemicals Ltd. See* dicamba, MCPA, mecoprop.

1294 Campbell's Redlegor

2893
2,4-DB [94-82-6] + MCPA [94-74-6].
Translocated herbicide for cereal crops. *MTM AgroChemicals Ltd. See* 2,4-DB, MCPA.

1295 Clovacorn Extra

An emulsifiable concentrate containing 220 g 2,4-DB, 30 g linuron and 30 g MCPA per liter; used to control weeds in undersown cereals and seedling grassland. *Farmers Crop Chemicals Ltd. See* 2,4-DB, linuron, MCPA.

1296 Dock-Ban

Dicamba + MCPA + mecoprop.
Soluble concentrate of 19.5 g dicamba, 245 g MCPA and 86.5 g mecoprop per liter; used for weed control in cereals and grassland. *Quadrangle Agrochemicals. See* Dicamba, MCPA, mecoprop.

1297 Docklene

Dicamba + MCPA + mecoprop.
Soluble concentrate of 336 g mecoprop, 84 g dicamba and 84 g MCPA per liter; used for weed control in cereals and grassland. *Schering Agrochemicals Ltd. See* Dicamba, MCPA, mecoprop.

1298 Farmon 2,4-DB Plus

2,4-DB + MCPA.
2,4-DB and MCPA; a herbicide for use with cereal crops. *Farm Protection Ltd.*

1299 Farmon 2,4-DP.CPA

Dichlorprop + MCPA.
Soluble concentrate of 358 g dichlorprop and 177 g MCPA per liter; used for control of weeds in cereals and grass. *Farm Protection Ltd.*

1300 Farmon MCPB Plus

MCPA + MCPB.
Soluble concentrate containing 41 g MCPA and 244 g MCPB per liter; for control of weeds in undersown cereals and grassland. *Farm Protection Ltd.*

1301 Greenmaster Extra

MCPA + Mecoprop.
Granules containing MCPA and mecoprop; for control of weeds in amenity and roadside grass. *Fisons plc. See* MCPA, mecoprop.

1302 Headland Relay

Dicamba, MCPA, mecoprop.
Mixture of dicamba, MCPA and mecoprop; used for weed control in cereals and grassland. *WBC Technology Ltd. See* dicamba, MCPA, mecoprop.

1303 Hemoxone

Dichlorprop + MCPA.
Soluble concentrate of 392 g dichlorprop and 210g MCPA per liter; used for control of weeds in cereals and grass. *ICI Chem & Polymers Ltd. See* dichlorprop, MCPA.

1304 Herrisol

Dicamba + MCPA + mecoprop.
A liquid containing 35.4% w/w dicamba (Banvel D), MCPA and mecoprop; used to control common chickweed, cleavers, knotgrass, redshank, scentless mayweed, corn spurrey and a wide range of other broadleaved weeds in cereals, grass crops and orchards. *Bayer plc. See* dicamba, MCPA, mecoprop.

1305 Hyprone

Dicamba + MCPA + mecoprop.
Soluble concentrate of 18.7 g dicamba, 100 g MCPA and 194 g mecoprop per liter, used for weed control in cereals and grassland. *Agrichem (International) Ltd.*

1306 Hysward

Dicamba + MCPA + mecoprop.
Mixture of dicamba, MCPA and mecoprop; used for weed control in cereals and grassland. *Agrichem (International) Ltd.*

1307 Legumex Extra
A solution concentrate containing 27 g benazolin, 237 g 2,4-DB and 42.3 g MCPA per liter; a post-emergence herbicide. *Schering Agrochemicals Ltd.*

1308 Lontrel Plus
Soluble concentrate containing 15 g clopyralid, 420 g dichlorprop and 175 g MCPA per liter; a translocated herbicide for use on cereals. *(Sold in UK for DowElanco.) ICI Chem & Polymers Ltd.*

1309 Mascot Super Selective
Soluble concentrate of 15 g dicamba, 100 g MCPA and 200 g mecoprop per liter; used for weed control in cereals and grassland. *Rigby Taylor Ltd. See* mecoprop, dicamba, MCPA.

1310 Meganox Plus
An emulsifiable concentrate containing 100 g aminotriazole, 250 g atrazine and 100 g MCPA per liter; used for total weed control in non crop areas. *Agri-Technics Ltd. See* aminotriazole, atrazine, MCPA.

1311 Mephetol Extra
Soluble concentrate of 20.8 g dicamba, 333 g dichloprop and 200 g MCPA per liter; used for weed control in cereals and newly sown grass. BritAg Industries Ltd. *See* dicamba, dichloprop, MCPA.

1312 MSS 2,4-DB + MCPA
2,4-DB [94-75-7] and MCPA [94-74-6]; translocated herbicide applied to cereals and undersown clovers. *Mirfield Sales Services Ltd.*

1313 MSS 2,4-DP + MCPA
Mixture of dichlorprop and MCPA; used for weed control in cereals and turf. *Mirfield Sales Services Ltd.*

1314 MSS Mircam Plus
Soluble concentrate of 19.5 g dicamba, 245 g MCPA and 86.5 g mecoprop per liter; used for weed control in cereals and grassland. *Mirfield Sales Services Ltd. See* dicamba, MCPA, mecoprop.

1315 Paddox
Herbicide containing the sodium/potassium salts of MCPA, mecoprop, and dicamba. *ICI Chem & Polymers Ltd. See* mecoprop, dicamba.

1316 Paddox
Soluble concentrate of 18 g dicamba, 237 g MCPA, and 80 g mecoprop per liter; used for weed control in cereals and grassland. *Farm Protection Ltd. See* mecoprop, dicamba.

1317 Pasturol Plus
Soluble concentrate of 25 g dicamba, 200 g MCPA, and 400 g mecoprop per liter; broad-spectrum herbicide used for weed control in cereals and grassland. *Farmers Crop Chemicals Ltd. See* mecoprop.

1318 Pulsar
A solution concentrate containing 200 g bentazone and 200 g MCPA per liter; a post-emergence herbicide. *BASF plc.*

1319 Quadban
Soluble concentrate of 19.5 g dicamba, 245 g MCPA and 86.5 g mecoprop per liter; used for weed control in cereals and grassland. *Quadrangle Agrochemicals.*

1320 Selectrol
Soluble concentrate of 18 g dicamba, 252 g MCPA and 84 g mecoprop per liter; used for weed control in cereals and grassland. *R P Adams Ltd. See* dicamba, MCPA, mecoprop.

1321 Seritox
Mixture of dichlorprop and MCPA; used for weed control in cereals and turf. Rhone-Poulenc Crop Protection Ltd. *See* dichlorprop and MCPA.

1322 Serramix CDA
An emulsifiable concentrate containing 100 g aminotriazole, 200 g atrazine and 100 g MCPA per liter; used for total weed control in non crop areas. *Denoon CDS; Powaspray (CDA) Ltd. See* aminotriazole, atrazine, MCPA.

1323 Setter 33
A solution concentrate containing 50 g benazolin, 237 g 2,4-DB and 43 g MCPA per liter; a post-emergence herbicide. *DowElanco Ltd. See* benazolin, 2,4-DB, 43 g MCPA.

1324 Springcorn Extra
Soluble concentrate of 18 g dicamba, 360 g MCPA and 160 g mecoprop per liter; used for weed control in cereals and grassland. *Farmers Crop Chemicals Ltd. See* dicamba, MCPA, mecoprop.

1325 Tetralex Plus
Soluble concentrate of 18 g dicamba, 252 g MCPA and 84 g mecoprop per liter; used for weed control in cereals and grassland. *Shell UK. See* dicamba, MCPA, mecoprop.

1326 Tetroxone
Selective weedkiller containing bromoxynil, dichlorprop, ioxynil and MCPA as potassium salts. *ICI Chem & Polymers Ltd.*

1327 Triagran®

Bentazon, dichorprop, MCPA; post-emergence herbicide for control of broad-leaved weeds in winter and spring cereals. *BASF AG. See* bentazon, dichorprop, MCPA.

1328 Tribute

Mixture of dicamba, MCPA and mecoprop; used for weed control in cereals and grassland. *Chipman Ltd. See* dicamba, MCPA, mecoprop.

1329 Trifolex-Tra

Soluble concentrate containing 34 g MCPA and 216 g MCPB per liter; for control of weeds in undersown cereals and grassland. *Shell UK. See* MCPA, MCPB.

1330 Tritox

An aqueous concentrate containing MCPA, mecoprop, and dicamba; a selective herbicide for the control of broad-leaved weeds in turf. *Fisons plc, Horticultural Div. See* MCPA, mecoprop, dicamba.

1331 Tropotox Plus

Soluble concentrate containing 37.5 g MCPA and 262.5 g MCPB per liter; for control of weeds in undersown cereals and grassland. *Rhône-Poulenc Crop Protection Ltd. See* MCPA, MCPB.

Herbicides containing MCPB

1332 Acumen

Bentazone + MCPA + MCPB.
Post-emergence contact and translocated herbicide for undersown cereals. *BASF plc.*

1333 Farmon MCPB Plus

MCPA + MCPB.
Soluble concentrate containing 41 g MCPA and 244 g MCPB per liter; for control of weeds in undersown cereals and grassland. *Farm Protection Ltd.*

1334 Hedonal®

2,4-D MCPA + Dichlorprop + MCPP.
Range of herbicides containing 2,4-D MCPA, dichlorprop, MCPP either alone or in combinations; growth regulator herbicide used for control of weeds in cereals. *Bayer AG.*

1335 Trifolex-Tra

Soluble concentrate containing 34 g MCPA and 216 g MCPB per liter; for control of weeds in undersown cereals and grassland. *Shell UK. See* MCPA, MCPB.

1336 Tropotox Plus

Soluble concentrate containing 37.5 g MCPA and 262.5 g MCPB per liter; for control of weeds in undersown cereals and grassland. *Rhône-Poulenc Crop Protection Ltd. See* MCPA, MCPB.

Herbicides containing Mecoprop

1337 Banlene Plus

Soluble concentrate of 18g dicamba, 252g MCPA and 84g mecoprop per liter; used for weed control in cereals and grassland. *Schering Agrochemicals Ltd. See* dicamba, MCPA, mecoprop.

1338 Bellclo

Soluble concentrate containing 250g 2,4-DB and 53 g mecoprop per liter; a translocated herbicide. *MTM Agrochemicals Ltd. See* 2,4-DB, mecoprop.

1339 BH CMPP/2,4-D

2,4-D + mecoprop.
Soluble concentrate containing 116g 2,4-D and 250g mecoprop per liter; used to control weeds in grassland. *Rhône-Poulenc/Agri. See* 2,4-D, mecoprop.

1340 Boxolon

Herbicide containing clopyralid, bromoxynil and mecoprop. *ICI Chem & Polymers Ltd. See* clopyralid, bromoxynil, mecoprop.

1341 Campbell's Field Marshal, Grassland Herbicide

Mixture of dicamba, MCPA and mecoprop; used for weed control in cereals and grassland. *MTM AgroChemicals Ltd. See* dicamba, MCPA, mecoprop.

1342 Cleaval

cyanazine + mecoprop.
Suspension concentrate containing 60 g cyanazine and 400 g mecoprop per liter; a contact herbicide. *Shell UK. See* cyanazine, mecoprop.

1343 Di-Farmon

Dicamba-mecoprop.
Soluble concentrate of 21 g dicamba and 319g mecoprop per liter; used for weed control in cereals and grassland. *ICI Chem & Polymers Ltd; Farm Protection Ltd. See* dicamba, mecoprop.

1344 Dock-Ban

Dicamba + MCPA + mecoprop.
Soluble concentrate of 19.5 g dicamba, 245 g MCPA and 86.5 g mecoprop per liter; used for weed control in cereals and grassland. *Quadrangle Agrochemicals. See* dicamba, MCPA, mecoprop.

1345 Docklene

Dicamba + MCPA + mecoprop.
Soluble concentrate of 336 g mecoprop, 84 g dicamba and 84 g MCPA per liter; used for weed

control in cereals and grassland. *Schering Agrochemicals Ltd. See* dicamba, MCPA, mecoprop.

1346   Endox
3090

Dicamba + mecoprop.

Soluble concentrate of 112 g dicamba [1918-00-9] and 265 g mecoprop 7085-19-0] per liter; herbicide used for weed control in cereals and grassland. *Farmers Crop Chemicals Ltd. See* dicamba, mecoprop.

1347   Farmon Condox
Dicamba + mecoprop.

Soluble concentrate of 112 g dicamba and 265 g mecoprop per liter; used for weed control in cereals and grassland. *Farm Protection Ltd. See* dicamba, mecoprop.

1348   Fettel
Dicamba + mecoprop + triclopyr.

Emulsifiable concentrate of 78g dicamba, 130g mecoprop, and 72g triclopyr per liter; used for weed control in cereals and grassland. *Farm Protection Ltd. See* dicamba, mecoprop, triclopyr.

1349   Foxstar
Bifenox + isoproturon + mecoprop.

A liquid formulation containing 107 g bifenox, 286 g isoproturon and 143 g mecoprop per liter as a suspension concentrate; a post-emergence herbicide for the control of weeds in winter cereals. *Rhône-Poulenc Crop Protection Ltd. See* bifenox, isoproturon, mecoprop.

1350   Green Up Feed and Weed Plus Mosskiller
Dry powder containing NPK 8:4:4 plus 2,4-D, mecoprop and ferrous sulfate; combined fertilizer, weed and mosskiller. *Vitax Ltd.*

1351   Green Up Lawn Feed 'n Weed Plus Moss Killer
Dichlorophen + mecoprop + dichlorprop + dicamba + benazolin.

Liquid concentrate containing NK 11:4 plus dichlorophen, mecoprop, dichlorprop, dicamba and benazolin; combined feed, weed and mosskiller. *Vitax Ltd.*

1352   Green Up Lawn Spot Weedkiller
2,4-D + Mecoprop.

Trigger spray pack containing 2,4-D [1702-17-6] and mecoprop [7085-19-0]; selective spot weedkiller for lawn areas. *Vitax Ltd.*

1353   Greenmaster Extra
MCPA + Mecoprop.

Granules containing MCPA and mecoprop; for control of weeds in amenity and roadside grass. *Fisons plc. See* MCPA, mecoprop.

1354   Harrier
Mecoprop-3,6-dichloropicolinic acid + ioxynil.

Weedkiller containing mecoprop, 3,6-dichloropicolinic acid and ioxynil as potassium salts. *ICI Chem & Polymers Ltd. See* mecoprop, 3,6-dichloropicolinic acid, ioxynil.

1355   Headland Relay
Dicamba + MCPA + mecoprop.

Mixture of dicamba, MCPA and mecoprop; used for weed control in cereals and grassland. *WBC Technology Ltd. See* dicamba, MCPA, mecoprop.

1356   Herrifex DS
mecoprop.

A liquid containing 587.5 g per liter mecoprop as the potassium salt; used to control cleavers, common chickweed and a wide range of other broad-leaved weeds in cereals, sports turf, grass seed crops and apple and pear orchards. *Bayer plc. See* mecoprop.

1357   Herrisol
Dicamba + MCPA + mecoprop.

A liquid containing 35.4% w/w dicamba (Banvel D), MCPA and mecoprop; used to control common chickweed, cleavers, knotgrass, redshank, scentless mayweed, corn spurrey and a wide range of other broadleaved weeds in cereals, grass crops and orchards. *Bayer plc. See* dicamba, MCPA, mecoprop.

1358   Hyban
Dicamba + mecoprop.

Soluble concentrate of 18.7 g dicamba and 300 g mecoprop per liter; used for weed control in cereals and grassland. *Agrichem (International) Ltd. See* dicamba, mecoprop.

1359   Hygrass
Dicamba + mecoprop.

Soluble concentrate of 18.7 g dicamba and 300 g mecoprop per liter; used for weed control in cereals and grassland. *Agrichem (International) Ltd. See* dicamba, mecoprop.

1360   Hyprone
Dicamba + MCPA + mecoprop.

Soluble concentrate of 18.7 g dicamba, 100 g MCPA and 194 g mecoprop per liter, used for weed control in cereals and grassland. *Agrichem (International) Ltd. See* dicamba, MCPA, mecoprop.

1361   Hysward
Dicamba + MCPA + mecoprop.

Mixture of dicamba, MCPA and mecoprop; used

for weed control in cereals and grassland. *Agrichem (International) Ltd. See* dicamba, MCPA, mecoprop.

1362   Iotox
Ioxynil + mecoprop.
Soluble concentrate containing 72 g ioxynil and 214 g mecoprop per liter; used for weed control in turf. *Rhône-Poulenc Environmental Prods. Ltd. See* ioxynil, mecoprop.

1363   Jaguar
Benazolin + bromoxynil + ioxynil + mecoprop.
An emulsifiable concentrate containing 22.2 g benazolin, 55.6 g bromoxynil, 27.8 g ioxynil and 413 g mecoprop per liter; a post-emergence herbicide for cereal crops and grass. *Schering Agrochemicals Ltd. See* benazolin, bromoxynil, ioxynil, mecoprop.

1364   Mascot Selective
2,4-D + mecoprop.
Soluble concentrate containing 60 g 2,4-D and 200 g mecoprop per liter; used to control weeds in grassland. *Rigby Taylor Ltd. See* mecoprop.

1365   Mascot Super Selective
Soluble concentrate of 15 g dicamba, 100 g MCPA and 200 g mecoprop per liter; used for weed control in cereals and grassland. *Rigby Taylor Ltd. See* mecoprop, dicamba, MCPA.

1366   MSS CMPP/DP
Soluble concentrate of 520 g dichlorprop and 130 g mecoprop per liter; used for control of weeds in barley, wheat and oats. *Mirfield Sales Services Ltd. See* dichlorprop, mecoprop.

1367   MSS Mircam
Soluble concentrate of 18.7 g dicamba and 300 g mecoprop per liter; used for weed control in cereals and grassland. *Mirfield Sales Services Ltd. See* dicamba, mecoprop.

1368   MSS Mircam Plus
Soluble concentrate of 19.5 g dicamba, 245 g MCPA and 86.5 g mecoprop per liter; used for weed control in cereals and grassland. *Mirfield Sales Services Ltd. See* dicamba, MCPA, mecoprop.

1369   Musketeer
Suspension concentrate containing 50 g ioxynil, 250 g isoproturon and 180 g mecoprop per liter; used for control of weeds in wheat and barley. *Hoechst UK. See* oxynil, isoproturon, mecoprop.

1370   New Formulation SBK Brushwood Killer
Liquid concentrate containing 2,4-D, mecoprop

and dicamba; selective herbicide for use on coarse and woody weeds. *Vitax Ltd. See* 2,4-D, mecoprop, dicamba.

1371   Paddox
Herbicide containing the sodium/potassium salts of MCPA, mecoprop, and dicamba. *ICI Chem & Polymers Ltd. See* mecoprop, dicamba.

1372   Paddox
Soluble concentrate of 18 g dicamba, 237 g MCPA, and 80 g mecoprop per liter; used for weed control in cereals and grassland. *Farm Protection Ltd. See* mecoprop, dicamba.

1373   Pasturol Plus
Soluble concentrate of 25 g dicamba, 200 g MCPA, and 400 g mecoprop per liter; broad-spectrum herbicide used for weed control in cereals and grassland. *Farmers Crop Chemicals Ltd. See* mecoprop.

1374   Post-Kite
Suspension concentrate containing 50 g ioxynil, 250 g isoproturon and 180 g mecoprop per liter; used for control of weeds in wheat and barley. *Schering Agro-chemicals Ltd. See* ioxynil, isoproturon, mecoprop.

1375   Quadban
Soluble concentrate of 19.5 g dicamba, 245 g MCPA and 86.5 g mecoprop per liter; used for weed control in cereals and grassland. *Quadrangle Agrochemicals. See* dicamba, MCPA, mecoprop.

1376   Selectrol
Soluble concentrate of 18 g dicamba, 252 g MCPA and 84 g mecoprop per liter; used for weed control in cereals and grassland. *R P Adams Ltd. See* dicamba, MCPA, mecoprop.

1377   Select-Trol
Soluble concentrate containing 6.6% w/w 2,4-D and 250 g mecoprop; used to control weeds in grassland. *Chemsearch (UK) Ltd. See* 2,4-D, mecoprop.

1378   Seloxone
Soluble concentrate containing 15 g clopyralid and 510 g mecoprop per liter; a translocated herbicide for cereals and grassland. *ICI Chem & Polymers Ltd. See* clopyralid, mecoprop.

1379   Springcorn Extra
Soluble concentrate of 18 g dicamba, 360 g MCPA and 160 g mecoprop per liter; used for weed control in cereals and grassland. *Farmers Crop Chemicals Ltd.*

1380   Supertox
Mixture of 2,4-D and mecoprop; used to control weeds in grassland. *Rhône-Poulenc/Agri. See* 2,4-D, mecoprop.

1381   Swipe 560 EC
An emulsifiable concentrate containing 56 g bromoxynil, 56 g Ioxynil and 448 g mecoprop per liter; a post-emergence contact herbicide for cereal crops. *Ciba-Geigy Agrochemicals. See* bromoxynil, Ioxynil, mecoprop.

1382   Sydex
Soluble concentrate containing 125 g 2,4-D and 250 g mecoprop per liter; used to control weeds in grassland. *Synchemicals Ltd. See* 2,4-D, 250 g mecoprop.

1383   Synox
Soluble concentrate containing 75 g ioxynil and 225 g mecoprop per liter; used for weed control in turf. *Synchemicals Ltd. See* ioxynil, mecoprop.

1384   Terset
Bromoxynil + ioxynil + isoproturon + mecoprop; a contact herbicide for use in cereal crops. *Rhône-Poulenc Crop Protection Ltd.*

1385   Tetralex Plus
Soluble concentrate of 18 g dicamba, 252 g MCPA and 84 g mecoprop per liter; used for weed control in cereals and grassland. *Shell UK. See* dicamba, MCPA, mecoprop.

1386   Tribute
Mixture of dicamba, MCPA and mecoprop; used for weed control in cereals and grassland. *Chipman Ltd. See* dicamba, MCPA, mecoprop.

1387   Tritox
An aqueous concentrate containing MCPA, mecoprop, and dicamba; a selective herbicide for the control of broad-leaved weeds in turf. *Fisons plc, Horticultural Div. See* MCPA, mecoprop, dicamba.

1388   Verdone 2
Contains mecoprop and 2,4-D; selective lawn weedkiller. ICI Garden Products.

1389   Verdone CDA
94-75-7;7058-19-0   2802,5666   202-361-1
$C_8H_6Cl_2O_3;C_{10}H_{11}ClO_3$
2,4-D + Mecoprop.
[2,4-D]:   (2,4-dichlorophenoxy)acetic acid; Hedonal; Trinoxol; [Mecoprop]: (±)-2-(4-chloro-2-methylphenoxy)-propanoic acid;   (±)-2-[(4-chloro-o-tolyl)oxy]-propionic acid; mechlorprop; MCPP; CMPP; RD 4593; Astix CMPP; IsoCornox; Compitox;   Compitox   Plus;   Proponex-Plus.

Emulsifiable concentrate containing 6.7% 2,4-D, and 13.3% mecoprop; used to control weeds in grassland. [2,4-D]: mp = 138°; $bp_{0.4}$ = 160°; [Mecoprop]: mp = 95-96°; $[\alpha]_D^{25}$ = +19 (alcohol). *ICI Agrochemicals Professional Products.*

1390   Weed and Brushkiller
Emulsifiable concentrate containing 144 g 2,4-D, 32 g dicamba and 144 g mecoprop per liter; an herbicide to control perennial and woody weeds. *Synchemicals Ltd. See* 2,4-D, dicamba, mecoprop.

1391   Zennapron
Mixture of 2,4-D and mecoprop. Used to control weeds in grassland. *BP Oil Ltd.*

Herbicides containing Methabenzthiazuron

1392   Glytex
Isoxaben + methabenzthiazuron.
Mixture of isoxaben and methabenzthiazuron; soil-acting herbicide for winter cereals. *Bayer plc. See* isoxaben, methabenzthiazuron.

1393   Ustinex®
Products with combinations of different herbicidal compounds   (aminotriazole,   diuron, methabenzthiazuron, phenoxies); herbicides used for control of weeds on paths, open spaces, parks and sports grounds. *Bayer AG. See* aminotriazole, diuron, methabenzthiazuron.

Herbicides containing Monolinuron

1394   Gramonol
Paraquat + monolinuron.
Weedkiller containing paraquat [1910-42-5] and monolinuron [330-55-2]. *ICI Chem & Polymers Ltd. See* paraquat, monolinuron.

1395   Gramonol 5
Paraquat + monolinuron.
Suspension   concentrate   containing   154   g monolinuron [330-55-2] and 110 g paraquat [1910-42-5] per liter; for control of annual dicotyledons in cereals. *Hoechst UK.*

1396   Gramonol Five
Paraquat [1910-42-5] + monolinuron [330-55-2].
Suspension   concentrate   containing   154   g monolinuron and 110 g paraquat per liter; for control of annual dicotyledons in cereals. *ICI Agrochemicals. See* paraquat, monolinuron.

Herbicides containing Paraquat

1397   Dexuron
Diuron + paraquat.
Suspension concentrate containing 300 g diuron and 100 g paraquat per liter; used to control

weeds around trees and shrubs. *Chipman Ltd. See* diuron, paraquat.

1398   Dukatalon
Diquat + paraquat.
Mixture of diquat and paraquat; herbicide. *Makhteshim Chemical Works Ltd. See* diquat, paraquat.

1399   Farmon PDQ
Diquat + paraquat.
Soluble concentrate of 80 g diquat and 120 g paraquat per liter; used for weed control in field crops. *Farm Protection Ltd. See* diquat, paraquat.

1400   Gramixel
Paraquat + diuron.
Herbicide containing paraquat [1910-42-5] and diuron [330-54-1]. *ICI Chem & Polymers Ltd. See* paraquat, diuron.

1401   Gramonol
Paraquat + monolinuron.
Weedkiller containing paraquat [1910-42-5] and monolinuron [330-55-2]. *ICI Chem & Polymers Ltd. See* paraquat, monolinuron.

1402   Gramonol 5
Paraquat + monolinuron.
Suspension concentrate containing 154 g monolinuron [330-55-2] and 110 g paraquat [1910-42-5] per liter; for control of annual dicotyledons in cereals. *Hoechst UK. See* paraquat, monolinuron.

1403   Gramonol Five
Paraquat [1910-42-5] + monolinuron [330-55-2].
Suspension concentrate containing 154 g monolinuron and 110 g paraquat per liter; for control of annual dicotyledons in cereals. *ICI Agrochemicals. See* paraquat, monolinuron.

1404   Gramuron
Paraquat + diuron.
Herbicide containing paraquat [1910-42-5] and diuron [330-54-1]. *ICI Chem & Polymers Ltd. See* paraquat, diuron.

1405   Parable
Soluble concentrate of 100 g diquat and 100 g paraquat per liter; used for weed control in field crops. *ICI Agrochemicals. See* diquat, paraquat.

1406   Paracol
Herbicide containing diuron and paraquat. *ICI Chem & Polymers Ltd. See* diuron, paraquat.

1407   Pathclear
Contains aminotriazole, diquat, paraquat and simazine; long-acting weedkiller for paths, drives, and patios. *ICI Garden Products. See* aminotriazole, diquat, paraquat , simazine.

1408   Soltair
Mixture of diquat, paraquat, and simazine; total herbicide. *ICI Chem & Polymers Ltd. See* diquat, paraquat, simazine.

1409   Totacol
Herbicide based on diuron and paraquatermary. *ICI Chem & Polymers Ltd. See* diuron, paraquat.

1410   Weedol
Weedkiller containing paraquat [1910-42-5] and diquat [85-00-7] for gardeners. *ICI Garden Products. See* paraquat, diquat.

Herbicides containing Pendimethalin

1411   Encore
Isoproturon + pendimethalin.
Suspension concentrate containing 125 g isoproturon and 250 g pendimethalin per liter; used for annual weed control in winter wheat, rye and barley. *Cyanamid of Great Britain Ltd. See* isoproturon, pendimethalin.

1412   Trump
Isoproturon + pendimethalin.
Suspension concentrate containing 236 g isoproturon and 236 g pendimethalin per liter; used for annual weed control in winter wheat, rye, and barley. *Cyanamid of Great Britain Ltd. See* isoproturon, pendimethalin.

Herbicides containing Picloram

1413   Atladox HI
Soluble concentrate containing 240 g 2,4-D and 65 g picloram per liter; used to control weeds in non crop grass and grass verges. *Chipman Ltd. See* 2,4-D, picloram.

1414   Hydon
Bromacil + picloram.
Bromacil and picloram; used for total weed control in non crop areas. *Chipman Ltd.*

1415   Primatol AP
Triazine, picloram total herbicide. *Ciba plc.*

1416   Sprint
Emulsifiable concentrate of 375 g fenpropimorph [67306-03-0] and 225 g prochloraz [67747-09-5] per liter; used for mildew control in cereals. *Schering Agrochemicals Ltd.*

Herbicides containing Prometryn

1417   Peaweed
Suspension concentrate containing 152 g prometryn and 304 g terbutryn per liter; for weed control in peas, beans, and potatoes. *Pan Britannica Industries Ltd. See* prometryn, terbutryn.

1418   Spudweed
Suspension concentrate containing 152 g prometryn and 304 g terbutryn per liter; used for weed control in peas, beans, and potatoes. *Pan Britannica Industries Ltd. See* prometryn, terbutryn.

Herbicides containing Propachlor

1419   Ashlade CP
Suspension concentrate containing 86 g chloridazon and 400 g propachlor per liter. A residual herbicide for use on beet crops. *Ashlade Formulations Ltd. See* chloridazon, propachlor.

1420   Decimate
Chlorthal + dimethyl + propachlor.
Suspension concentrate containing 225 g chlorthal-dimethyl and 216 g propachlor per liter; an herbicide for use on brassicas and onions. *Fermenta ASC Europe Ltd. See* chlorathal, dimethyl, propachlor.

1421   Sorgan
Propachlor + propazine; herbicide. *Agan Chemical Manufacturers Ltd. See* propachlor, propazine.

Herbicides containing Propham

1422   Atlas Electrum
Suspension concentrate containing 200 g chloridazon, 30 g chlorpropham, 20 g fenuron and 120 g propham per liter; a residual herbicide for beet crops. *Atlas Interlates Ltd. See* chloridazon, chlorpropham, fenuron, propham.

1423   Barrier
Suspension concentrate containing 300 g chloridazon, 30 g fenuron and 170g propham per liter; a residual herbicide for beet crops. *Truchem Ltd. See* chloridazon, fenuron, propham.

1424   Matrikerb
Mixture of clopyralid and propyzamide; a soil and leaf herbicide for winter oilseed rape. *Pan Britannica Industries Ltd. See* clopyralid and propyzamide.

1425   MSS Sugar Beet Herbicide
A suspension concentrate containing 37.5 g chlorpropham, 25 g fenuron and 150 g propham per liter; an herbicide for use on beet crops.

Mirfield Sales Services Ltd. See chlorpropham, fenuron, propham.

1426   Premalox
Herbicide and growth regulator based upon propham. *May & Baker Ltd. See* propham.

1427   Truchem Quintex
Emulsifiable concentrate containing 30 g chlorpropham, 30 g fenuron and 130 g propham per liter; an herbicide for beet crops. *MTM Agrochemicals Ltd. See* chlorpropham, fenuron, propham.

Herbicides containing Simazine

1428   Atlacide Extra
Atrazine-sodium chlorate; used for total weed control in non crop areas. *Chipman Ltd. See* atrazine, sodium chlorate.

1429   Aventox SC
Simazine-trietazine; herbicide used with peas and beans. *Dow Elanco Ltd. See* simazine, trietazine.

1430   Banweed-S
A suspension concentrate containing napropamide and simazine; soil-applied residual for control of annual grasses and annual broad-leaved weeds in field and container grown nursery stock. *Fisons plc, Horticultural Div. See* napropamide, simazine.

1431   Boroflow S/ATA
A suspension concentrate containing 160 g aminotriazole and 270 g simazine per liter; used for total weed control in non crop areas and fruit orchards. *ABM Chemicals Ltd. See* aminotriazole, simazine.

1432   CDA Simflow Plus
A suspension concentrate containing 100 g aminotriazole [61-82-5] and 300 g simazine [122-34-9] per liter; used for total weed control in non crop areas and fruit orchards. *Rhône-Poulenc Environmental Prods.* Ltd. *See* aminotriazole, simazine.

1433   Clearway
aminotriazole-simazine.
A suspension concentrate containing 100 g aminotriazole [61-82-5] and 300 g simazine [122-34-9] per liter; used for total weed control in non crop areas and fruit orchards. *Rhône-Poulenc Environmental Prods.* Ltd. *See* aminotriazole, simazine.

1434   Herbazin Plus
Simazine + aminotriazole.
A wettable powder containing simazine and aminotriazole; a quick acting herbicide for control

of existing weeds with long term persistence. *Fisons plc, Horticultural Div. See* simazine, aminotriazole.

1435 Herbazin Plus SC

Amitrole + simazine.

A suspension concentrate containing 180 g aminotriazole [61-82-5] and 300 g simazine [122-34-9] per liter; used for total weed control in non crop areas and fruit orchards. *Fisons plc. See* amitrole, simazine.

1436 Hermes

Metoxuron + simazine.

Suspension concentrate containing 480 g metoxuron and 30 g simazine per liter; post-emergence herbicide for winter cereals. *Atlas Interlates Ltd.*

1437 Hytrol

Aminotriazole + 2,4-D + diuron + simazine.

Aminotriazole, 2,4-D, diuron and simazine; used for total weed control in non crop areas. *Farmers Crop Chemicals Ltd.*

1438 Hytrol

2,4-D + diuron + amitrole + simazine.

2,4-D, diuron, amitrole, simazine; total weedkiller with more than one seasons resistance; for use on garden paths and other noncrop areas. *Agrichem (International) Ltd.*

1439 Mascot Highway

Aminotriazole [61-82-5] and simazine [122-34-9]; used for total weed control in non crop areas. *Rigby Taylor Ltd. See* aminotriazole, simazine.

1440 Mascot Ultrasonic

Glyphosate + simazine.

Glyphosate + simazine; a translocated and residual herbicide for the control of grasses and weeds. *Rigby Taylor Ltd. See* glyphosate, simazine.

1441 MSS Simazine/Aminotriazole 43FL

A suspension concentrate containing 155 g [122-34-9] per liter; used for total weed control in non crop areas and fruit orchards. *Mirfield Sales Services Ltd. See* aminotriazole, simazine.

1442 Pathclear

Contains aminotriazole, diquat, paraquat and simazine; long-acting weedkiller for paths, drives, and patios. *ICI Garden Products. See* aminotriazole, diquat, paraquat , simazine.

1443 Pearson's solution

A solution of dried sodium arsenate 0.1% - 1.0% .

1444 Remtal

Trietazine [1912-26-1] and simazine [122-34-9]; selective herbicide for peas and beans. *Schering Agrochemicals Ltd. See* trietazine, simazine.

1445 Simazol

Active ingredients: azolan (aminotriazole) plus simanex (simazine); multipurpose herbicidal mixture which eradicates a wide spectrum of established weeds, while preventing further weed germination for extended periods. *Agan Chemical Manufacturers Ltd. See* simazine, aminotriazole.

1446 Simfix

Granular mixture of diuron and simazine; used for control of weeds in woody crops and noncrop areas. *Rhône-Poulenc Environmental Prods. Ltd.*

1447 Simflow Plus

A suspension concentrate containing 100 g aminotriazole [61-82-5] and 300 g simazine [122-34-9] per liter; used for total weed control in non crop areas and fruit orchards. *Rhône-Poulenc Environmental Prods. Ltd. See* aminotriazole, simazine.

1448 Soltair

Mixture of diquat, paraquat, and simazine; total herbicide. *ICI Chem & Polymers Ltd. See* diquat, paraquat, simazine.

1449 Synchemicals Total Weed Killer

A suspension concentrate containing 53 g aminotriazole [61-82-5] and 110 g simazine [122-34-9] per liter; used for total weed control in non crop areas and fruit orchards. *Synchemicals Ltd. See* aminotriazole, simazine.

1450 Syntox Total Weed Killer

A suspension concentrate containing 53 g aminotriazole [61-82-5] and 100 g simazine [122-34-9] per liter; used for total weed control in non crop areas and fruit orchards. *Syntex Manufacturing Ltd. See* aminotriazole, simazine.

1451 Weedkill

Aminotriazole + 2,4-D + diuron + simazine; used for total weed control in non crop areas. *Dermaglen Ltd.*

Herbicides containing Terbuthylazine

1452 Gardoprim A 500FW

Atrazine + terbuthylazine.

A suspension concentrate containing 100 g atrazine and 400 g terbuthylazine per liter; used for total weed control in forestry plantations. *Ciba-Geigy Agrochemicals.*

1453 Opogard 500

Suspension concentrate containing 150 g terbuthylazine and 350 g terbutryn per liter; weed germination inhibitor. *Ciba-Geigy Agrochemicals.* *See* terbuthylazine, terbutryn.

Herbicides containing Terbutryn

1454 Opera

Suspension concentrate containing 150 g terbutryn and 200 g trifluralin per liter; for weed control in winter cereals. *Tripart Farm Chemicals Ltd.* *See* terbutryn, trifluralin.

1455 Opogard 500

Suspension concentrate containing 150 g terbuthylazine and 350 g terbutryn per liter; weed germination inhibitor. *Ciba-Geigy Agrochemicals.* *See* terbuthylazine, terbutryn.

1456 Peaweed

Suspension concentrate containing 152 g prometryn and 304 g terbutryn per liter; for weed control in peas, beans and potatoes. *Pan Britannica Industries Ltd. See* prometryn, terbutryn.

1457 Primatol AP

Triazine, picloram total herbicide. *Ciba plc. See* triazine, picloram.

1458 Senate

Suspension concentrate containing 250 g terbutryn [886-50-0] and 250 g trietazine [1912-26-1] per liter; herbicide for weed control in potatoes, peas and field beans. *Schering Agrochemicals Ltd. See* terbutryn, trietazine.

1459 Spudweed

Suspension concentrate containing 152 g prometryn and 304 g terbutryn per liter; used for weed control in peas, beans and potatoes. *Pan Britannica Industries Itd. See* prometryn, terbutryn.

1460 Summit

Suspension concentrate containing 150 g terbutryn and 200 g trifluralin per liter; for weed control in winter cereals. *Ashlade Formulations Ltd. See* terbutryn, trifluralin.

1461 Tempo

Herbicide containing linuron and terbutryn. *ICI Chem & Polymers Ltd. See* linuron, terbutryn.

1462 Tempo

Suspension concentrate containing 150 g linuron and 150 g terbutryn per liter; used for control of annual dicotyledons in potatoes. *Farm Protection Ltd. See* linuron, terbutryn.

1463 Terbalin

Active ingredients; terbutryn plus triflurex; selective pre-emergence herbicidal mixture. *Agan Chemical Manufacturers Ltd. See* terbutryn, triflurex.

Herbicides containing Triclopyr

1464 Broadshot

Emulsifiable concentrate containing 200 g 2,4-D, 85 g dicamba and 65 g triclopyr per liter; an herbicide to control perennial and woody weeds. *Shell UK. See* 2,4-D, dicamba, triclopyr.

1465 Fettel

Dicamba + mecoprop + triclopyr. Emulsifiable concentrate of 78g dicamba, 130g mecoprop, and 72g triclopyr per liter; used for weed control in cereals and grassland. *Farm Protection Ltd. See* dicamba, mecoprop, triclopyr.

1466 Grazon 90

Clopyralid + triclopyr. Emulsifiable concentrate containing 60 g clopyralid and 240 g triclopyr per liter; used for treatment of perennial weeds in grassland. *Dow Elanco Ltd. See* clopyralid, triclopyr.

Herbicides containing Trietazine

1467 Aventox SC

Simazine-trietazine; herbicide used with peas and beans. *DowElanco Ltd. See* simazine, trietazine.

1468 Pre-Empt

Emulsifiable concentrate containing 46 g linuron, 54 g trietazine and 200 g trifluralin per liter; herbicide for winter cereals. *Schering Agrochemicals Ltd. See* linuron, trietazine, trifluralin.

1469 Remtal

Trietazine [1912-26-1] and simazine [122-34-9]; selective herbicide for peas and beans. *Schering Agrochemicals Ltd. See* trietazine, simazine.

1470 Senate

Suspension concentrate containing 250 g terbutryn [886-50-0] and 250 g trietazine [1912-26-1] per liter; herbicide for weed control in potatoes, peas and field beans. *Schering Agrochemicals Ltd. See* terbutryn, trietazine.

Herbicides containing Trifluralin

1471 Ardent
Suspension concentrate containing 40 g diflufenican and 400 g trifluralin per liter; used for control of weeds in winter cereals. *Embetec Crop Protection Ltd. See* diflufenican, trifluralin.

1472 Autumn Kite
Emulsifiable concentrate of 300 g isoproturon [34123-59-6] and 200 g trifluralin [1582-09-8] per liter; used for control of annual grasses in winter wheat and barley. *Schering Agrochemicals Ltd. See* isoproturon, trifluralin.

1473 Campbell's Trifluron
Mixture of linuron [330-55-2] and trifluralin [1582-09-8]; herbicide for winter cereals. *MTM AgroChemicals Ltd. See* linuron, trifluralin.

1474 Chandor
Emulsifiable concentrate containing 120 g linuron and 240 g trifluralin per liter; herbicide for winter cereals. *DowElanco Ltd. See* linuron, trifluralin.

1475 Flint
Linuron + trifluralin.
Emulsifiable concentrate containing 120 g linuron and 240 g trifluralin per liter; herbicide for winter cereals. *Ashlade Formulations Ltd.*

1476 Janus
Linuron + trifluralin.
Liquid mixture of linuron [330-55-2] and trifluralin [1982-09-8]; herbicide for winter cereals. *Atlas Interlates Ltd.*

1477 Marksman
Mixture of linuron and trifluralin; herbicide for winter cereals. *Farmers Crop Chemicals Ltd.*

1478 Onslaught
Suspension concentrate containing 160 g linuron [330-55-2] and 320 g trifluralin [1582-09-8] per liter; herbicide for winter cereals. *Quadrangle Agrochemicals. See* linuron, trifluralin.

1479 Opera
Suspension concentrate containing 150 g terbutryn and 200 g trifluralin per liter; for weed control in winter cereals. *Tripart Farm Chemicals Ltd. See* terbutryn, trifluralin.

1480 Pre-Empt
Emulsifiable concentrate containing 46 g linuron, 54 g trietazine and 200 g trifluralin per liter; herbicide for winter cereals. *Schering Agrochemicals Ltd. See* linuron, trietazine, trifluralin.

1481 Solo
Emulsifiable concentrate containing 1250 g linuron [330-55-2] and 240 g trifluralin [1582-09-8] per liter; herbicide for winter cereals. *MTM Agrochemicals Ltd. See* linuron, trifluralin.

1482 Summit
Suspension concentrate containing 150 g terbutryn and 200 g trifluralin per liter; for weed control in winter cereals. *Ashlade Formulations Ltd. See* terbutryn, trifluralin.

1483 Trifarmon
Suspension concentrate containing 160 g linuron [330-55-2] and 320 g trifluralin [1582-09-8] per liter; herbicide for winter cereals. *Farm Protection Ltd. See* linuron, trifluralin.

1484 Tri-Farmon
Herbicide containing linuron [330-55-2] and trifluralin [1582-09-8]. *ICI Chem & Polymers Ltd. See* linuron, trifluralin.

Miscellaneous Herbicide Combinations

1485 Ancrack
Naptalam + dinitro.
Naptalam plus dinitro; pre-emergent herbicide for use on peanuts and soybeans. *Draxel Chemical Co. See* naptalam, dinitro.

1486 Ashlade D-Moss
Chloroxuron + ferrous sulfate.
Mixture of chloroxuron and ferrous sulfate; a lawn sand herbicide to control mosses in turf. *Ashlade Formulations Ltd. See* chloroxuron, ferrous sulfate.

1487 Atlas Brown
Chlorpropham + pentanochlor.
Emulsifiable concentrate containing 150 g chlorpropham and 100 g pentanochlor per liter, a residual herbicide. *Atlas Interlates Ltd. See* chlorpropham, pentanochlor.

1488 Atlas Red
Chlorpropham + cresylic acid + fenuron.
Suspension concentrate containing 200 g chlorpropham, cresylic acid and 50 g fenuron per liter; an herbicide for use on ornamentals and vegetables. *Atlas Interlates Ltd.*

1489 Atramet Combi
Atranex + ametrex.
Active ingredients; atranex plus ametrex, ready formulated mixture of atrazine plus ametryne for use as a selective pre-and post-emergence herbicide. *Agan Chemical Manufacturers Ltd.*

1490   Banweed-S

napropamide + simazine.
A suspension concentrate containing napropamide and simazine; soil-applied residual for control of annual grasses and annual broad-leaved weeds in field and container grown nursery stock. *Fisons plc, Horticultural Div.* See napropamide, simazine.

1491   Bromotrill

Bromoxynil octanoate + 2,6-dibromo-4-cyanophenyl octanoate.
Active ingredients: bromoxynil octanoate and 2,6-dibromo-4-cyanophenyl octanoate; selective postemergence control of a wide range of annual broadleaf weeds in winter and spring cereals and in com. *Agan Chemical manufacturers Ltd.*

1492   Bronco

2-Chloro-2',6'- diethyl-N-(methoxymethyl) acetanilide + glyphosate isopropylamine salt.
Active ingredients are 2.6 lb of 2-chloro-2',6'-diethyl-N-(methoxymethyl) acetanilide and 1.4 lb of the isopropylamine salt of glyphosate; herbicide for no-till farming. *Monsanto Co.*

1493   Bronox

Linuron + tritrazine.
Wettable powder containing linuron and tritrazine; used for control of annual dicotyledons and annual grasses in potatoes and nursery stock. *Schering Agrochemicals Ltd.*

1494   Croak

Fluometuron + MSMA.
Fluometuron plus MSMA; herbicide for post-emergence control of broadleaf and grass weeds in cotton. *Draxel Chemical Company.* See fluometuron, MSMA.

1496   Decimate

Chlorthal + dimethyl + propachlor.
Suspension concentrate containing 225 g chlorthal-dimethyl and 216 g propachlor per liter; an herbicide for use on brassicas and onions. *Fermenta ASC Europe Ltd.* See chlorthal-dimethyl, propachlor.

1497   Dicurane Duo 495FW

Bifenox + chlorotoluron.
A liquid formulation containing 106 g bifenox and 389 g chlorotoluron per liter as a suspension concentrate; a residual herbicide for the control of weeds in winter wheat. *Ciba-Geigy Agrochemicals.* See bifenox, chlorotoluron.

1498   Diurol

Azolan + diurex.
Active ingredients: azolan plus diurex; multlpurpose herbicidal mixture which eradicates a wide spectrum of established weeds while preventing further weed germination for extended periods. *Agan Chemical Manufacturers Ltd.* See azolan, diurex.

1499   DUK-880

Lenacil + phenmedipham.
Lenacil and phenmedipham; used for control of annual dicotyledons in sugar beet. *DuPont UK.* See lenacil, phenmedipham.

1500   Fenocil

Bromacil + pentachlorophenol.
Bromacil and pentachlorophenol; used for total weed control in noncrop areas. *Chemsearch (UK) Ltd.* See bromacil, pentachlorophenol.

1501   Galaxy®

Bentazon.
acifluorphen. For post-emergence control of annual broadleaf weeds in soybeans and peanuts. *BASF AG.* See bentazon.

1502   Harrier

Mecoprop-3,6-dichloropicolinic acid + ioxynil.
Weedkiller containing mecoprop, 3,6-dichloropicolinic acid and ioxynil as potassium salts. *ICI Chem & Polymers Ltd.*

1503   Hermes

Metoxuron + simazine.
Suspension concentrate containing 480 g metoxuron and 30 g simazine per liter; post-emergence herbicide for winter cereals. *Atlas Interlates Ltd.* See metoxuron, simazine.

1504   Holdfast D

Dicamba + paclobutrazol.
Soluble concentrate of 25 g dicamba and 250 g paclobutrazol per liter; used for weed control in cereals and grassland. *ICI Agrochemicals.*

1505   Komeen®

Copper + ethylenediamine.
Copper-ethylenediamine complex; aquatic herbicide for hydrilla control in golf course, ornamental, and fish ponds, potable water reservoirs, fresh water lakes, fish hatcheries. *Griffin.*

1506   Oracle®

Isoproturon and metsulfuron + methyl.
Isoproturon and metsulfuron-methyl; for annual weed control in wheat and barley. *DuPont UK.*

1507   Pinnacle

Imazamethabenz + methyl and isoproturon.
Mixture of imazamethabenz-methyl and isoproturon; used for control of weed grasses in winter cereals. *Cyanamid of Great Britain Ltd.*

1508   Pyradur®
Chloridazon + metolachlor.
Chloridazon and metolachlor; pre-emergence herbicide for control of grasses and broad-leaved weeds in sugar and fodder beet crops. *BASF AG.*

1509   Ronstar TX
Carbetamide + oxadiazon.
Carbetamide [16118-49-3] and oxadiazon [19666-30-9]; a residual herbicide for pre-weed emergence control for container grown nursery plants. *Hortichem Ltd.*

1511   Sprint
Fenpropimorph + prochloraz.
Emulsifiable concentrate of 375 g fenpropimorph [67306-03-0] and 225 g prochloraz [67747-09-5] per liter; used for mildew control in cereals. *Schering Agrochemicals Ltd.* See fenpropimorph, prochloraz.

1512   Terbalin
Triflurex + terbutryn.
Active ingredients; terbutryn plus triflurex; selective pre-emergence herbicidal mixture. *Agan Chemical Manufacturers Ltd.* See terbutryn, triflurex.

1513   Uradex
Uragan + diurex.
Active ingredients; diurex plus uragan; selective pre-emergence herbicide mixture. *Agan Chemical Manufacturers Ltd.* See diurex, uragan.

1514   Verdict
Haloxyfop.
A line of herbicides based primarily on haloxyfop. *Dow UK.*

1515   Volunteered
Di-1-p-menthene + dalapon.
Mixture of dalapon and di-1-p-menthene; a grass weed herbicide. *Mandops (UK) Ltd.* See dalapon, di-1-p-menthene.

Herbicide Products with Uncertain Composition

1516   Actril
Selective weed killer. *May & Baker Ltd.*

1517   Actril S
Bromoxynil + dichloroprop + oxynil + MCPA.
Broad spectrum, post emergence contact and translocated herbicide. *Rhône-Poulenc Crop Protection Ltd.*

1518   Actrilawn
Selective weedkiller. *May & Baker Ltd.*

1519   Agritox
Selective weedkiller. *May & Baker Ltd.*

1520   Arbogard
Herbicides. *ICI Chem & Polymers Ltd.*

1521   Astrol
Bromoxynil + ioxynil + isoproturon.
A contact herbicide for cereal crops. *Embetec Crop Protection Ltd.*

1522   Atlas Brown
Herbicide used to protect vegetable crops. *Allied Colloids Ltd.*

1523   Atlas Gold
Herbicide used to protect sugar beet. *Allied Colloids Ltd.*

1524   Atlas Pink C
Pesticides for vegetables. Herbicide. *Allied Colloids Ltd.*

1525   Atlas Protrum® K
Pesticides for sugar beet. Herbicide. *Allied Colloids Ltd.*

1526   Atlas Red
Pesticides for vegetables. Herbicide. *Allied Colloids Ltd.*

1527   Atlas Solan
Pesticides for vegetables. Herbicide. *Allied Colloids Ltd.*

1528   Atlas Steward®
Herbicide for cereals. *Allied Colloids Ltd.*

1529   Atlas Tecgran
Pesticides for vegetables. Herbicide. *Allied Colloids Ltd.*

1530   Atlas Total A, Total S
Pesticides for forestry and amenity products. Herbicide. *Allied Colloids Ltd.*

1531   Atlavar
Atrazine + 2,4-D + sodium chlorate.
Weed control in non crop areas. *Chipman Ltd.*

1532   Atragan
Herbicide. *Agan Chemical Manufacturers Ltd.*

1533   Axall
Herbicide. *May & Baker Ltd.*

1534   Banwee
Residual herbicide. *Fisons plc, Horticultural Div.*

1535   Borascu
Borax decahydrate.
General nonselective weedkillers. *Borax Consolidated Ltd.*

1536   BR Destral
Aminotriazole + bromacil + diuron.
Used for total weed control in noncrop areas and on railway tracks. *ABM Chemicals Ltd.*

1537   Brittox
Herbicide. *May & Baker Ltd.*

1538   Brush-B-Gon
Brush killer. *Monsanto (Solaris).*

1539   Cambilene
3206
A selective weedkiller. *Fisons plc, Horticultural Div. See* dicamba.

1540   Carbon 4E
A liquid formulation containing 48% trichloropyrester; controls woody weeds and perennial broad-leaved weeds in forestry and noncropped areas. *Burts & Harvey.*

1541   Centrifugal Syrup
A selective weedkiller. *A H Marks & Co Ltd.*

1542   Composibor
Destral BR. Nonselective herbicides; Weedkilling. *Borax Europe Ltd.*

1543   Defolia
Defoliant for hops. *Murphy Chemical Co Ltd.*

1544   Dicotox
A selective weedkiller. *May & Baker Ltd.*

1545   Dinamene
Selective weedkiller. *Murphy Chemical Co Ltd.*

1546   Di-On
Herbicide. *Agan Chemical Manufactures Ltd.*

1547   Doff
Range of insecticides and herbicides; for horticultural/household use. *Doff Portland Ltd.*

1548   Doublet
Herbicide. *May & Baker Ltd.*

1549   Duplosan®
Translocated herbicide for cereals and grassland. *BASF plc.*

1550   Evergreen
Lawn fertilizer combined with selective weedkiller. *Fisons plc, Horticultural Div.*

1551   Farmon
Range of liquid herbicides of different formulations. *ICI Chem & Polymers Ltd.*

1552   Fernesta
A selective weed killer. *Plant Protection.*

1553   Finesse®
Weed killer. *DuPont UK.*

1554   Fisons P.C.P.
A weedkiller. *Fisons plc, Horticultural Div.*

1555   Forlay
Selective weed killers. *Murphy Chemical Co Ltd.*

1556   Frenokone
A selective weedkiller. Plant Protection.

1557   Glean® TP
Bromoxynil + chlorsulfuron + ioxynil.
Herbicide mixture for weed control in cereals. *DuPont UK. See* bromoxynil, chlorsulfuron, ioxynil.

1558   Glydus
Herbicide. *Agan Chemical Manufacturers Ltd.*

1559   Gramazine
Herbicide. *ICI Chem & Polymers Ltd.*

1560   Graminon® Plus
bentazon + isoproturon + dichlorprop.
For post-emergence control of grasses and broadleaf weeds in winter wheat and winter barley. *BASF AG. See* bentazon, soproturon, dichlorprop.

1561   Graslam
Herbicide. *May & Baker Ltd.*

1562   Groundhog
Aminotriazole + diquat + paraquat + simazine.
Used for total weed control in non crop areas. *ICI Agrochemicals Professional Products; ICI Chem. & Polymers Ltd.*

1563   Herbazin Special
Amitrole + atrazine + 2,4-D.
Used for total weed control in non crop areas.
Fisons plc.

1564   Herbrak®
Herbicide for controlling the grass and broad
leaved weeds occurring in sugar and fodder beet;
predrilling, pre-emergence or post-emergence
application. *Bayer AG.*

1565   Hi-bor
Destral. Nonselective herbicides; For weedkilling.
Borax Europe Ltd.

1566   Hotspur
Herbicide for broad leafed weeds in cereals. *ICI
Chem & Polymers Ltd.*

1567   Iotril
Ioxynil octanoate.
4-cyano-2,6-diiodophenyl octanoate. Selective
post-emergence herbicide which controls a wide
range of annual broadleaf weeds in cereals,
onions, leeks, and sugar cane. *Agan Chemical
Manufactures Ltd. See* ioxynil [1689-83-4].

1568   Iotrilex
Ioxynil octanoate.
Herbicide. *Agan Chemical Manufactures Ltd. See*
ioxynil [1689-83-4].

1569   Iso-Cornox
Selective weedkillers. *The Boots Co plc.*

1570   Iso-Planotox
Selective weedkiller. *May & Baker Ltd.*

1571   Ivorin-Profalon
Herbicide for beans and potatoes. Hoechst UK.

1572   Ivosit
Selective contact herbicide. *Hoechst UK.*

1573   Kilmet
Selective weedkillers. *May & Baker Ltd.*

1574   Kilnet
Selective weedkiller. *May & Baker Ltd.*

1575   Kleenup
Weed and grass Killer. *Monsanto (Solaris).*

1576   Laddok®
Bentazon.
Atrazine. Herbicide for selective postemergence
control of broadleaf weeds. BASF AG.

1577   Lasso
2-Chloro-2',6'-diethyl-N-
(methoxymethyl)acetanilide.
Pre-emergence herbicide. *Monsanto Co.*

1578   Lawn Plus
Fertilizer and selective weedkiller. *ICI Chem &
Polymers Ltd.*

1579   Lawn Spot Weeder
Selective weedkiller aerosol. *Fisons plc,
Horticultural Div.*

1580   Lawn Weedkiller
Selective weedkiller. *Murphy Chemical Co Ltd.*

1581   Lawnsman
Range of lawn aids for garden use as fertilizers,
weedkillers, and a spreader. *ICI Garden Products.*

1582   Lawnsman Mosskiller
Mosskiller for lawns. *ICI Chem & Polymers Ltd.*

1583   Lawnsman Weed and Feed
Fertilizer combined with selective weedkiller. *ICI
Chem & Polymers Ltd.*

1584   Ley Cornox
Selective weedkillers. *The Boots Co plc.*

1585   Mag-40
Liquid chemical defoliant of cotton and a
desiccant in silverskin onion. *Makhteshim
Chemical Works Ltd.*

1586   Matrikerb
Clopyralid + propyzamide.
A post-emergence herbicide for use on winter
oilseed rape. *Rohm & Haas UK. See* clopyralid
and propyzamide.

1587   Mayclene
Selective herbicide.
Engelhard Technologies Ltd.

1588   Mephetol
Herbicides. *ICI Chem & Polymers Ltd.*

1589   Merpelan AZ
Wettable powder formulation for pre-emergent
control of weeds. *Bayer AG.*

1590   Mofix
Oxime/triazine herbicide. *Ciba plc.*

1591   Morto
Weedkiller and destroyer of potato haulm (stems
and stalks). *Murphy Chemical Co Ltd.*

1592   New Legumex

Selective weedkillers. *Fisons plc, Horticultural Div*.

1593   New Murbetex

Pre-emergence herbicide. *Murphy Chemical Co Ltd*.

1594   New Verdone

Selective weedkiller. *Plant Protection*.

1595   Orchard Herbide

Amitrole [61-82-5] + diuron [330-54-1].
Used for total weed control in fruit orchards. *Hoechst UK. See* amitrole, diuron.

1596   Oxytril

Selective weedkiller. *May & Baker Ltd*.

1597   Oxytril P

Selective herbicide. *Murphy Chemical Co Ltd*.

1598   Path Gun+se

Ready-for-use herbicide spray. *ICI Chem & Polymers Ltd*.

1599   Pre-Kite

Selective herbicide. *Schering Agrochemicals Ltd*.

1600   Pyracur® FL

Chloridazon + metolachlor.
Pre-emergence herbicide for control of grasses and broadleaf weeds in sugar beet and fodder beet. *BASF AG. See* chloridazon, metolachlor.

1601   Pyradex® T

Chloridazon + triallate.
Pre-plant incorporated herbicide for control of broadleaf weeds and grasses in sugar and fodder beet. *BASF AG. See* chloridazon, triallate.

1602   Rancho®

Herbicide; controls grass weeds and some broad-leaved weeds in irrigated rice crops. *Bayer AG*.

1603   Rassamix CDA

Amitrole + atrazine + diuron.
A mixture of herbicides for weed control. *Denoon CDS. See* amitrole, atrazine, diuron.

1604   Rassapron

Amitrole + atrazine + diuron.
A mixture of herbicides for weed control. *BP Oil Ltd. See* amitrole, atrazine, diuron.

1605   RegalStar

Herbicide to control crabgrass, goosegrass and other annual weeds; applied in early spring to turfgrass and cultivated nursery fields prior to weed seed germination. *Regal Chemical Company*.

1606   Seritox

Selective weedkiller. *May & Baker Ltd*.

1607   Sodium Pentachlorphenate

C_6Cl_5NaO
Pentachlorophenol sodium salt.
sodium pentachlorophenate; sodium pentachlorophenoxide; Santobrite; Dowicide G. Insecticide, fungicide and non-selective contact herbicide. Used in control of termites, as a wood preservative, a pre-harvest defoliant and a general pre-emergence herbicide. Soluble in H_2O; [free pentachlorophenol]: LD_{50} (rat orl) = 210 mg/kg.

1608   Solubor DF

BR Destral. Heavy-duty nonselective herbicide; used for weedkilling. *Borax Europe Ltd*.

1609   Springclene 2

Selective herbicide. *Schering Agrochemicals Ltd*.

1610   Starane 2

Selective post-emergence herbicide. *Murphy Chemical Co Ltd. See* fluroxypyr 1-methylheptyl ester.

1611   Storm®

Bentazon + acifluorfen.
For post-emergence control of broadleaf weeds in soybeans and peanuts. *BASF AG. See* acifluorfen.

1612   Super Weedex

Total weedkiller. *Murphy Chemical Co Ltd*.

1613   Supersevtox

Selective herbicide. *Schering Agrochemicals Ltd*.

1614   Supertox

Selective weedkiller. *May & Baker Ltd*.

1615   Talent

Herbicide. *May & Baker Ltd*.

1616   Terraklene

Herbicide. *ICI Chem & Polymers Ltd*.

1617   Tim-bor Professional

Borocil. Nonselective weedkiller. *Borax Europe Ltd*.

1618   Trigger

Herbicide. *May & Baker Ltd*.

1619   Triox

Vegetation killer. *Monsanto (Solaris)*.

1620 Tumbleweed Gel

Weedkiller for spot application. *Murphy Chemical Co Ltd.*

1621 Twin-Tak

Herbicide. *May & Baker Ltd.*

1622 U 46®

Used for weed control in cereals, maize, sugar cane, rice, grassland, forestry, perennial crops, tree crops. *BASF AG. See* 2,4-D, MCPA, dichlorprop, mecoprop.

1623 Vamitox

Herbicide. *May & Baker Ltd.*

1624 Vanitox

Selective weed killer. *May & Baker Ltd.*

1625 Vectal

Selective and total herbicide. *Schering Agrochemicals Ltd.*

1626 Weedone® DPC

2,4-DP + 2,4-D.

Post-emergent herbicide for control of annual and perennial broadleaf weeds on golf courses and other ornamental turf area. *Rhône-Poulenc/Ag; W.A. Cleary. See* 2,4-DP, 2,4-D.

Insecticides

1627 Arsenic Trioxide

1327-53-3   844   215-481-4

As_4O_6

Arsenic (III) oxide.

Arsenic oxide;Arsenous trioxide; arsenous acid; arsenous oxide; arsenic sesquioxide; White Arsenic; Diarsenic Trioxide; Crude Arsenic; Arsenic (white); Arsenious oxide; Arsenic (III) trioxide; Arsenous anhydride; arsenite; arsenolite; arsenous acid anhydride; arsenous oxide anhydride; arsodent; claudelite; claudetite; Arsenic oxide (3); Arsenic oxide (As_2O_3); Arsenic sesquioxide (As_2O_3); Arsenicum album; Diarsonic trioxide; Diarsenic oxide. Pigments, ceramic enamels, aniline colors, decolorizing agent in glass, insecticide, rodenticide, herbicide, sheep and cattle dip, hide preservative, wood preservative, preparation of other arsenic compounds. mp = 315°; bp = 465°; soluble in H_2O, dil HCl, alkali hydroxide or carbonate solns; insoluble in EtOH, $CHCl_3$, Et_2O; LD_{50} (rat orl) = 1.46 mg/kg. *Atomergic Chemetals; Noah Chem.; Outokumpu Oy; Transene.*

1628 Calcium Arsenate

7778-44-1   1686   231-904-5

$As_2Ca_3O_8$

Calcium arsenate.

Tricalcium arsenate; Pencal; Calcars. Used as insecticide, molluscicide Slightly soluble in H_2O and dilute acids; LD_{50} (rat orl) = 298 mg/kg. *Mechema Chemicals Ltd.*

1629 Calcium Oxide

1305-78-8   1733   215-138-9

CaO

Lime.

quicklime; calx; CaO. Inorganic oxide; refractory, sewage treatment, insecticides, fungicides, manufacture of steel and aluminum; flotation of nonferrous ores; manufacture of glass, paper, Ca salts; in drilling fluids, lubricants; laboratory. mp = 2572°; bp = 2850°; d = 3.32-3.35. *Cerac; GE; Hüls Am.; Mallinckrodt; Pfizer; U.S. Gypsum.*

1630 Carbaryl

63-25-2   1831   200-555-0

$C_{12}H_{11}NO_2$

1-Naphthyl methylcarbamate.

1-Naphthalenol, methylcarbamate; carbamic acid, methyl-, 1-naphthyl ester; α-naphthalenyl methylcarbamate; α-naphthyl N-methylcarbamate; Atoxan; Bercema NMC 50; Caprolin; Carbamic acid, N-methyl-1-naphthyl-; Carbamic acid, N-methyl,1-naphthyl ester; Carbamine; Carbaril; Carbarilo; Carbarilum; Carbatox; Carbatox 75; Carbatox-60; Carbavur; Carpolin; Compound 7744; Denapon; Dicarbam; Dyna-Carbyl; Tomado; Carylderm; Microcarb; Murvin 85; Thinsec; Tornado; Wasp Destroyer. Used as an insecticide. Used in lice infestation. A suspension concentrate containing 450 g carbaryl per liter. contact insecticide and fruit thinner for apples. mp = 142°; d = 1.232; soluble in H_2O (40 mg/l), freely soluble in most organic solvents; LD_{50} (rat orl) = 230 mg/kg. *Napp Laboratories Ltd; ICI AgroChemicals Professional Products; ICI Agrochemicals.*

1631 Chlorfenvinphos

470-90-6   2137   207-432-0

$C_{12}H_{14}Cl_3O_4P$

Phosphoric acid 2-chloro-1-(2,4-dichlorophenyl)-ethenyl diethyl ester.

2-chloro-1-(2,4-dichlorophenyl)-vinyl diethyl phosphate; β-2-Chloro-1-(2',4'-dichlorophenyl)-vinyl diethyl phosphate; 2,4-dichloro-α-(chloromethylene)diethyl benzyl alcohol, phosphate; Birlan; Birlane; Birlane 10G; C 8949; C-10015; Clofenvinphos; chlorofenvinphos; Clofenvineosum; Clofenvinfos; CVP; SD-7859; Compd. 4072; Dermaton; Sapecron; Steladone; Supona. Insecticide and acaricide with contact and stomach action. Used in seed treatment or soil application for control of root flies, rootworms, fruit flies, Colorado beetles, etc. mp = -22 to -16°; bp$_{0.05}$ = 167-170°; soluble in H$_2$O (145 mg/l), miscible with organic solvents; LD$_{50}$ (rat orl) = 9.6 mg/kg.

1632   Confidor
105827-78-9
C$_9$H$_{10}$ClN$_5$O$_2$
1-[(6-Chloro-3-pyridinyl)methyl]-4,5-dihydro-N-nitro-1H-imidazol-2-amine.
1-(6-Chloro-3-pyridiylmethyl)-N-nitroimidazoli-din-2-ylideneamine; Admire; Gaucho; NTN 33 893; Imidacloprid. Systemic insecticide with contact and stomach action; applied as a foliar or soil treatment especially against sucking pests (virus vectors), e.g., aphids, thrips, whiteflies and leafhoppers on rice, potatoes, vegetables, cotton, tobacco, citrus, pome, stone fruit and other crops; also against some biting insects such as rice water weevil, Colorado potato beetle, wireworm, frit fly, beet fly, onion fly and citrus and apple leaf miners. mp = 136.5-144°; soluble in H$_2$O (0.51 g/l); LD$_{50}$ (rat orl) = 450 mg/kg. *Bayer AG.*

1633   Cupric Acetate, Basic
52503-64-7   2691   257-974-7
Cupric subacetate.
complex with different cupric acetate:cupric hydroxide:water ratios; verdigris, green verdigris. Used as pigments, insecticides, fungicides, mold-preventatives.

1634   Cypermethrin
52315-07-8   2836   257-842-9
C$_{22}$H$_{19}$Cl$_2$NO$_3$
3-(2,2-Dichloroethenyl)-2,2-dimethylcyclopropanecarboxylic acid cyano(3-phenoxyphenyl) methyl ester.
Barricade, Cymbush; FMC 30980; FMC 45806; Imperator; Kafil; PP 383; Ripcord; WL 43467; Aimcocyper; Ammo; Arrivo; Basathrin; CCN52; Cymperator; Cynoff; Cyper; Cypercopal; Cyperguard; Cyperkill; Cyperscet; Cypertox; Demon; Fenom; Flectron; Fligene CI; Folcord; Halt; LE 79600; NRDC 149; Nurelle; Polytrin; Prevail; Ralothrin; Sherpa; Siperin; Stockade; Toppel; Ustaad; WL 43467. Non-systemic insecticide with contact and stomach action. Used to control wide range of insects, e.g. *lepidoptera, coleoptera, diptera* and *hemiptera*. mp = 60-80°; d^{20} = 1.25; insoluble in H$_2$O, soluble in organic solvents; LD$_{50}$ (rat orl) = 70 mg/kg.

1635   Diallyl Maleate
999-21-3   213-658-0
C$_{10}$H$_{12}$O$_4$
Polymers and copolymers, insecticides. *Aceto; Ashland.*

1636   Diazinon Liquid
333-41-5   3043   206-373-8
C$_{12}$H$_{21}$N$_2$O$_3$PS
O,O-Diethyl-O-(6-methyl-2-(1-methylethyl)-4-pyrimidinyl)phosophorothioate.
Spectracide; Dimpylate; Knox Out; Dianon; Gardentox; Kayazinon; G-24480; Basudin; Neocidol; Dipofene; Diazitol; Ag-500; Antigal; Dacutox; Dassitox; Dazzel; Diagran; Diaterr-fos; Diazajet; Diazide; Diazol; Diethyl Dimpylatum; Dizinon; Drawizon; Dyzol; Exodin; Fezudin; Flytrol; Galesan; Kayazol; Knox Out 2FM; Neocidol; Nipsan; Nucidol; Sarolex. Insecticide. mp = 120°; bp = 306°; d = 1.117; slightly soluble in H$_2$O (0.004 g/100 ml), more soluble in organic solvents; LD$_{50}$ (rat orl) = 76 mg/kg. *DowElanco Ltd.*

1637   Ethion
563-12-2   3782   209-242-3
C$_9$H$_{22}$O$_4$P$_2$S$_4$
O,O,O',O'-Tetraethyl-S,S'-methylene di(phosphorodithioate).
Ethanox; Ethiol; FMC 1240; Hylemox; Rhodiacide; Rhodocide; RP-Thion; Vegfru-Fosmite. Has both acaricidal and insecticidal properties; its acaricidal action is widely used in the abatement of cattle ticks; as a non-systemic insecticide it is used on citrus, deciduous fruits, tea, cotton and ornamental plants. mp = -12 to -15°; bp$_{0.3}$ = 164-165°; slightly soluble in H$_2$O, more soluble in organic solvents; LD$_{50}$ (rat orl) = 208 mg/kg. *A/S Cheminova.*

1638   Imidazole
288-32-4   4948   206-019-2
C$_3$H$_4$N$_2$
1,3-Diaza-2,4-cyclopentadiene.
Glyoxaline; glyoxalin; 1,3-diazole; iminazole; miazole; pyrro[b]monazole. Biological control of pests, especially fabric-feeding insects; contact insecticide in an oil spray. mp = 90-91°; soluble in H$_2$O, organic solvents; LD$_{50}$ (mus orl) = 1880 mg/kg. *Aldrich; BASF; Janssen Chimica; Penta Mfg.*

1639   Kafil
52645-53-1   7321   258-067-9
C$_{21}$H$_{20}$Cl$_2$O$_3$
Permethrin.
3-(2,2-Dichloroethenyl)-2,2-dimethylcyclopro-panecarboxylic acid (3-phenoxyphenyl)methyl ester; 3-(phenoxyphenyl)methyl (±)-cis,trans-3-(2,2-dichloroethenyl)-2,2-dimethylcyclopropane-carboxylate; m-phenoxybenzyl(±)-cis,trans-3-(2,2-dichlorovinyl)-2,2-dimethylcyclopropane-carboxylate: FMC-33297; NIA-33297; NRDC-143; PP-557; SBP-1513; S-3151; Fumite Permethrin; Ambush; Corsair; Dragnet; Ectiban; Eksmin; Nix; Pulvex; Pounce; Pynosect; Ridect PourOn; Cooper

Coopex; Darmycel Agarifume Smoke; Elimite®;; Kafil; Nippon Ant and Crawling Insect Killer; Nippon Ant Killer Powder; Nyppon Fly Killer Spray; Nippon Ready for Use Ant and Crawling Insect Killer; Nix Creme Rinse; Nix Delmal Cream; Perigen; Permasect; Picket; Quamilin; Turbair Permethrin. Smoke insecticide (active ingredient permethrin); for use in enclosed areas against whitefly and other pests of protected crops, cockroaches on stored produce, domestic insect pests. mp $\cong$ 35°; bp$_{0.005}$ = 220°; d^{20} = 1.190-1.272. *ICI Chem & Polymers Ltd.; Octavius Hunt Ltd; ICI Chem. & Polymers Ltd; The Welcome Foundation Ltd; darmycel UK; Allergan Inc; Vitax Ltd; Mitchell Cotts Chemicals Ltd; ICI Garden Products; Pan Britannica Industries Ltd.*

1640 Karate
68085-85-8   2827   268-450-2
C$_{23}$H$_{19}$ClF$_3$NO$_3$
λ-cyhalothrin.
3-(2-Chloro-3,3,3-trifluoro-1-propenyl)-2,2-dimethylcycloproanecarboxylic acid cyano(3-phenoxyphenyl)methyl ester; Grenade. Pyrethroid insecticide. *ICI Chem & Polymers Ltd.*

1641 Kaydox
106-46-7   3107   203-400-5
C$_6$H$_4$Cl$_2$
1,4-Dichlorobenzene paste.
p-Dichlorbenzene;   Paracide;   PDB; paradichlorobenzene; Para-zene; Di-chloricide; Paramoth. Insecticide Sublimes; mp = 53.5°; bp 174.12°; n$_D^{60}$ = 1.5285; LD$_{50}$ (rat orl) = 500 mg/kg. *Murphy Chemical Co Ltd.*

1642 Killgerm® Py-Kill W
7696-12-0   9362   231-711-6
C$_{19}$H$_{25}$NO$_4$
Tetramethrin.
2,2-Dimethyl-3-(2-methyl-1-propenyl)cyclo-propanecarboxylic acid (1,3,4,5,6,7-hexahydro-1,3-dioxo-2H-isoindol-2-yl)methyl ester; 2,2-dimethyl-3-(2-methylpropenyl)cyclopropane-carboxylic acid ester with N-(hydroxymethyl)-1-cyclohexene-1,2-dicarbox-imide;   N-(3,4,5,6-tetrahydrophthalimide)-methyl-*cis,trans*-chrys-anthemate;   N-(chrysanthemoxy-methyl)-1-cyclohexene-1,2-dicarboximide;   phthal-thrin; FMC-9260; SP-1103; Neo-Pynamin. Emulsifiable concentrate containing 22 g/l tetramethrin; for control of files in livestock houses. mp = 65-80°; d$_{20}^{20}$ = 1.108; n$_D^{21.5}$ = 1.5175. Killgerm Chemicals Ltd.

1643 Korlan
299-84-3   8415   206-082-6
C$_8$H$_8$Cl$_3$O$_3$PS
Ronnel.
Phosphorthioic acid, O,O-dimethyl O-(2,4,5-trichlorophenyl)ester; fenchlorphos; dimethyl trichlorophenyl thiophosphate; Trolene; Etrolene; Nankor; Viozene; Ectorl. Active ingredient: ronnel; insecticide used on cattle for the control of ticks, files, maggots, and lice. mp = 41°. *Dow UK.*

1644 Lorexane
58-89-9   5526   200-401-2
Preparations of γ benzene hexachloride (lindane); an insecticide; antiparasitic hair lotion. *ICI Chem & Polymers Ltd.*

1645 Mephosfolan
950-10-7   213-447-3
C$_8$H$_{16}$NO$_3$PS$_2$
(4-Methyl-1,3-dithiolan-2-ylidene)phosphoramidic acid diethyl ester.
phosphonodithioimidocarbonic   acid   cyclic propylene   P,P-diethyl   ester;   2-diethoxyphosphinylimino-4-methyl-1,3-dithiolane; cyclic propylene(diethoxyphosphinyl)dithioimidocarbonate; EI 47470; ENT 25,991; Cytrolane; Cytro-Lane. Emulsifiable concentrate containing 250 g/l mephosfolan; used for control of damsonhop aphid in hops. bp$_{0.001}$ = 120°; soluble in H$_2$O (57 mg/l), soluble in organic solvents; LD$_{50}$ (rat orl) = 8.9 mg/kg. *Cyanamid of Great Britain Ltd.*

1646 Omethoate
1113-02-6   214-197-8
C$_5$H$_{12}$NO$_4$PS
O,O-Dimethyl   S-[2-(methylamino)-2-oxoethyl] phosphorothioate.
dimethoate-met; Bay 45432; Folimat; S-6876; Omethoate. Systemic insecticide and acaricide with contact and stomach action. Used for control of spider mites, aphids, beetles, caterpillars, scale insects, thrips, suckers, fruit flies etc. in fruit and vegetable crops and in forestry. Dec 135°; d^{20} = 1.32; n$_D^{20}$ = 1.4987; soluble in H$_2$O, organic solvents; LD$_{50}$ (rat orl) = 50 mg/kg. *Bayer AG; Bayer plc; Bayer.*

1647 Propoxur
114-26-1   8022   204-043-8
C$_{11}$H$_{15}$NO$_3$
2-(1-Methylethoxy)phenyl methyl carbamate.
2-isopropoxyphenyl   methyl   carbamate;   o-isopropoxyphenyl methyl carbamate; 58 12 315; arporcarb; OMS 33; ENT 25671; Bay 39007; Baygon; Blattanex; Brifur; Invisi-Gard; Pillargon; Propogon; Rhoden; Sendran; Suncide; Tendex; Tugen; Unden; Undene; Undeen. Non-systemic insecticide used for treatment of sucking and biting insects, e.g., aphids, mealybugs, scales, leafhoppers, caterpillars on vegetables, pome and stone fruit, cocoa, rice, oil palms and other crops. Used for protection of flowers, fruits and mp = 84-87°; d = 1.12; soluble in H$_2$O (2 g/l), more soluble

in organic solvents; LD_{50} (rat orl) = 8350 mg/kg. *Bayer AG.*

1648   Rampart

1563-66-2   1851   216-353-0

$C_{12}H_{15}NO_3$

2,3-Dihydro-2,2-dimethyl-7-benzofuranol methylcarbamate.

7-Benzofuranol,   2,3-dihydro-2,2-dimethyl-, methylcarbamate; Carbamic acid, methyl-, 2,3-dihydro-2,2-dimethyl-7-benzofuranyl ester; BAY 70143; BAY 78537; C2292-59a; Carbamic acid, methyl-, 2,2-dimethyl-2,3-dihydrobenzofuran-7-yl ester; Carbamic acid, methyl-, 2,2-dimethyl-2,3-dihydro-7-benzofuranyl   ester;   Carbofuran; Carbofuran; Carbofuran; Carbofurane; Chinufur; Crisfuran; Curaterr; D 1221; 2,3-Dihydro-2,2-dimethylbenzofuran-7-yl methylcarbamate; 2,3-Dihydro-2,2-dimethylbenzo-furanyl-7-N-methyl-carbamate;   2,2-Dimethyl-7-coumaranyl   N-methylcarbamate;   2,2-Dimethyl-2,2-dihydro-benzofuranyl-7 N-methylcarbamate; ENT 27,164; FMC 10242; Furacarb; Furadan; Furadan 3g; Furadan 3G; Furadan 4f; Furadane; Furadan; Furadan 75 wp; Furodan; Karbofuranu; Me f248; Methyl carbamic acid 2,3-dihydro-2,2-dimethyl-7-benzofuranyl   ester;   NIA   10242;   Niagara 10242;OMS 864;Yaltox. A free flowing granule containing 5% w/w carbofuran; soil-applied insecticide for control of soil and seedling pests including cabbage root fly, cabbage stem weevil, flea beetle, cabbage stem flea beetle, early aphids in brassicas, turnip root fly, frit flies, millipedes, symphilids,   beet   leaf   miner,   springtails, wireworms, free living nematodes, potato cyst eelworm, carrot fly and carrot willow aphid. Cholinesterase inhibitor. mp = 150-153°; soluble in H_2O (700 mg/l); LD_{50} (rat orl) = 2 mg/kg. *Sipcam UK Ltd; Universal Crop Protection Ltd.; Bayer AG; Bayer plc.*

1649   Rapid

23103-98-2   7651   245-430-1

$C_{11}H_{18}N_4O_2$

Dimethylcarbamic acid 2-(dimethylamino)-5,6-dimethyl-4-pyrimidinyl ester.

Pirimicarb; 5,6-dimethyl-2-dimethylamino-4-dimethylcarbamoyloxypyrimidine; PP-062; EN-27766; Aphox; Fernos; Pirimor. Garden insecticide. mp = 91°; soluble in H_2O (2.7 g/l), organic solvents; LD_{50} (rat orl) = 147 mg/kg. *ICI Chem & Polymers Ltd.*

1650   Reldan 50

5598-13-0   2242   227-011-5

Chlorpyrifos-methyl.

An organophosphate insecticide for the treatment of pests in stored grain and oilseed rape. *DowElanco Ltd. See* chloropirifos-methyl.

1651   Taktic

33089-61-1   510   251-375-4

$C_{19}H_{23}N_3$

N-(2,4-Dimethylphenyl)-N-[[(2,4-dimethylphenyl)imino]methyl]-N-methylmethaniminamide.

N,N'[(methylimino)dimethylidyne]di-2,4-xylidine; N-methylbis(2,4-xylyliminomethyl)amine;   N,N-bis(2,4-xylyliminomethyl)methylamine; N-methyl-N'-2,4-xylyl-N-(N,2,4-xylylformimidoyl)-formamidine; 1,5-bis(2,4-dimethylphenyl)-3-methyl-1,3,5-triazapenta-1,4-diene; amitraz; amitraze; OMS 1820;ENT 27967; BTS 27419; Acadrex; Acarac; Azadieno; BAAM; Bumetran; Danicut; Ectodex; Edrizar; Istambul; Maitac; Mitac; Ovasyn; Topline; Triatix; Triatox; Tudy. Animal health insecticide. Non-systemic acaricide and insecticide with contact and respiratory action. Thought to interact with octopamine receptors in the nervous system, causing an increase in nervous activity. Used for control at all stages of mites, pear suckers, scale insects, mealy bugs, whitefly, aphids and lepidoptera. mp = 86-87°; d = 1.128; soluble in H_2O (1 mg/l), more soluble in organic solvents; LD_{50} (rat orl) = 800 mg/kg. *Schering Agrochemicals Ltd; Atabay; NOR-AM; Quimica Estrela.*

1652   Talon

2921-88-2   2242   220-864-4

$C_9H_{11}Cl_3NO_3PS$

Phosphorothioic acid O,O-diethyl O-(3,5,6-trichloro-2-pyridinyl) ester.

chlorpyrifos;   O,O-diethyl-O-3,5,6-trichloro-2-pyridyl   phosphorothioate;   chlorpyrifos-ethyl; Dowco 179; ENT-27311; Dursban; Lorsban; Pyrinex. Emulsifiable concentrate containing 228g chlorpyrifos per liter; broad-spectrum insecticide with many crop uses. Insecticides containing chlorpyrifos are used to control ticks on cattle, mosquitoes, and other insects. mp = 41-42°; insoluble in H_2O (2 ppm), very soluble in organic solvents; λ_m = 208, 230, 290 nm; LD_{50} (rat orl) 145 mg/kg. *Farmers Crop Chemicals Ltd.; Dow UK.*

1653   Tamaron®

10265-92-6   6014   233-606-0

$C_2H_8NO_2PS$

O,S-Dimethyl phosphoramidothioate.

Methamidophos;acephate-met; Bay 71628; Filitox; Monitor; Patrole; Pillaron; SRA 5172; Tam; Tamanox. Broad spectrum systemic insecticide with stomach and contact action used for treatment of sucking and biting insects and spider mites on a wide range of crops including cotton, tobacco, vegetables, potatoes, sugar beets. mp = 46°; d = 1.31; n_D^{40} = 1.5092; very soluble in H_2O (> 2 kg/l), less soluble in organic solvents; LD_{50} (rat orl) = 20 mg/kg. *Bayer AG.*

1654 Terbufos

13071-79-9   9301   235-963-8

$C_9H_{21}O_2PS_3$

S-t-butylthio-methyl-O,O-diethyl phosphorodithioate.

S-tert-butylthiomethyl O,O-diethyl phosphorodithioate; S-[[(1,1-dimethylethyl)thio]methyl] O,O-diethyl phosphorodithioate; ENT-27920; AC 92100; Aragran; Contraven; Counter; Cyanater; Plydax. A soli insecticide and nematicide with stomach and contact action. A cholinesterase inhibitor. Used for control of soil insects in vegetable and fruit crops. mp = -29°; $bp_{0.01 \ mm}$ = 69°; d_{24} = 1.105; n_D^{23} = 1.52; soluble in H_2O (4.5 mg/l), more soluble in organic solvents; LD_{50} (rat orl) = 1.6 mg/kg. *American Cyanamid.*

1655 Terracur® P

115-90-2   4042   204-114-3

$C_{11}H_{17}O_4PS_2$

O,O-Diethyl O-[4-(methylsulfinyl)phenyl] phosphorothioate.

O,O-diethyl O-4-methylsulfinylphenyl phosphorothioate; DMSP; OMS 37; ENT 24945; Bay 25141; Dasanit; S 767. Granular insecticide and nematicide with primarily contact action; for treatment of biting insects and nematodes. Cholinesterase inhibitor. Used for control of nematodes and soil-dwelling insects. $bp_{0.01}$ = 138-141°; d_{20} = 1.202; n_D^{25} = 1.540; soluble in H_2O (1.54 g/l), soluble in most organic solvents; LD_{50} (rat orl) = 10.5 mg/kg. *Bayer AG.*

1656 Terrathion

298-02-2   7486   206-052-2

$C_7H_{17}O_2PS_3$

O,O-Diethyl S-[(ethylthio)methyl]-phosphorodithioate.

AC 3911; Agrimet; Chim; Forate; Geomet; Granutox; Rampart; Thimet; Vegfru Foratox; Volphor. Consists of granules containing 10% w/w phorate; an organophosphorus insecticide. mp < -15°; $bp_{0.8}$ = 118-120°; d_{25} = 1.167; n_D^{25} = 1.5349; soluble in H_2O (50 mg/l), more soluble in organic solvents; LD_{59} (rat orl) = 3.7 mg/kg. *Farmers Crop Chemicals Ltd.*

1657 Thiabendazole

148-79-8   9426   205-725-8

$C_{10}H_7N_3S$

2-(4-Thiazolyl)-1H-benzimidazole.

Thiabendazole; benzimidazole, 2-(4-thiazolyl)-; Apl-Luster; Arbotect; Bioguard; Bovizole; Eprofil; Equizole; Lombristop; Mertec; Mertect; Mertect 160; Metasol TK 100; Mintesol; Mintezol; Minzolum; Mycozol; MK 360; Nemapan; Omnizole; Polival; Tbz; Tebuzate; Tecto; Tecto RPH; Tecto 10P; Tecto 40F; Tecto 60; Testo; Thiaben; Thiabendazol; Thiprazole; Tiabenda.

Systemic fungicide with protective and curative action. Absorbed by leaves and roots; used for control of fungus in vegetabls, fruits and cereals. mp = 300° (dec); soluble in H_2O (250 mg/l at pH 2-5), more soluble in organic solvents; LD_{50} (rat orl) = 2080 mg/kg. *MSD Agvet; Pennwalt; Duphar; Agrichem; BASF, Ciba-Geigy; Dow Elanco.*

1658 Thiodan

115-29-7   3614   204-079-4

$C_9H_6Cl_6O_3S$

6,7,8,9,10,10-Hexachloro-1,5,5a,6,9,9a-hexahydro-6,9-methano-2,4,3-benzodioxathiepin 3-oxide.

Endosulfan; benzoepin; OMS 750; ENT 23979; Beosit; Cyclodan; Chlortiepin; Devisulphan; Endocel; Endosol; FMC 5462; Hilda; Hoe 2671; Insectophene; Malix; Rasayansulfan; Thifor; Thimul; Thionex; Thiosulfan. Non-systemic insecticide and acaricide with contact and stomach action. Used in control of sucking, chewing and boring insects and mites on a wide variety of crops. The commercial product is a mixture of an α-isomer (mp = 108-110°) and a β-isomer (mp = 208-210°). Endosulfan is a chlorinated cyclic sulfurous acid ester having broad spectrum insecticidal activity of long-lasting effect. mp = 109°; $bp_{0.7}$ = 106°; d_{20} = 1.745; soluble in H_2O (0.32 mg/l), more soluble in organic solvents; LD_{50} (rat orl) = 70 mg/kg. *Hoechst UK; Makhteshim Chemical Works Ltd.*

1659 Thripstick®

52918-63-5   2934   258-256-6

$C_{22}H_{19}Br_2NO_3$

[1R-[1α(S*),3α]]-Cyano(3-phenoxyphenyl)methyl 3-(2,2-dibromomethenyl)-2,2-dimethylcyclopropanedicarboxylate.

Deltamethrin; deltamethrine; Butoflin; Butox; Cislin; Crackdown; Decis; Delsekte; K-Otek; K-Othrin; NRDC 161; RU 22974. A fast-acting non-systemic pyrethroid insecticide with contact and stomach action. Used to control many species of insect in many crops. Non-phytotoxic. mp = 98-101°; insoluble in H_2O (< 0.002 mg/l), very soluble in organic solvents; LD_{50} (rat orl) = 128 mg/kg. *Aquaspersions Ltd; Hoechst UK.*

1660 Vapona

62-73-7   3129   200-547-7

$C_4H_7Cl_2O_4P$

Dichlorvos.

Phosphoric acid2,2-dichloroethenyl dimethyl ester; phosphoric acid 2,2-dichlorovinyl dimethyl ester; O,O-dimethyl O-(2,2,-dichlorovinyl) phosphate; dichlorophos; dichlorovos; DDVP; SD 1750; Astrobot; Atgard; Canogard; Dedevap; Dichlorman; Divipan; Equigard; Equigel; Estrosol;

Herkol; Nogos; Nuvan; Task; Verdisol. A proprietary preparation of dichlorvos; an insecticide. $d_4^{25} = 1.415$; $bp_{20} = 140°$; $bp_{1.0} = 84°$; $bp_{0.5} = 72°$; $bp_{0.01} = 30°$; $n_D^{25} = 1.451$.

1661   Vassgro DSM

919-86-8   6129   213-052-6

Emulsifiable concentrate containing 500 g demeton-S-methyl per liter; systemic organophosphorus insecticide. *L W Vass (Agri) Ltd.*

1662   Volathion®

14816-18-3   238-887-3

$C_{12}H_{15}N_2O_3PS$

4-Ethoxy-7-phenyl-3,5-dioxa-6-aza-4-phosphaoct-6-ene-8-nitril 4-sulfide.

Phoxim; O,O-diethyl α-cyanobenzylideneamino-oxyphosphonothioate; 2-(diethoxyphosphinothio-yloxyimino)-2-phenylacetonitrile; phoxime; Bay 5621; Bay 77488; Baythion; Volaton. Foliage- and soil-applied insecticide for control of lepidopterous larvae, beetles and their larvae and locusts on a wide range of crops including cotton maize and vegetables. mp = 5-6°; $d^{20} = 1.176$; $n_D^{20} = 1.5405$; soluble in H_2O (7 mg/l), very soluble in organic solvents; LD_{50} (rat orl) = 1976-2170 mg/kg. *Bayer AG.*

1663   XL-All Insecticide

54-11-5   6611   200-193-3

$C_{10}H_{14}N_2$

3-(1-Methyl-2-pyrollidinyl)pyridine.

Nicotine; 1-methyl-2-(3-pyridiyl)pyrrolidine; β-pyridyl-α-N-methylpyrrolidine. Component of leaves of *Nictoiana tabacum* and *N. rustica*. An alkaloid insecticide. bp = 243-248°; d = 1.017; $n_D^{20} = 1.5265$; $[α]_D^{20} = -168±5°$ (c = 5, H_2O); miscible with H_2O, soluble in organic solvents. *Synchemicals Ltd.*

1664   Zineb

12122-67-7   10300   235-180-1

$(C_4H_6N_2S_4Zn)_x$

[[1,2-Ethanediylbis[carbamodithioato]](2-)]zinc.

zinc ethylenebis(dithiocarbamate) (polymeric); zinebe, Zidan; Aaphytora; Acuprex; Aspor; Dipher; Dithane Z 78; Diiner; Ditiozin; Enozin; Hexaphane; Kypzin; Lonacol; Parzate; Permilan; Phytox; Polyram-Z; Sepineb; Tiezene; Tritoftorol; Zinosan; Zinugec. Foliar fungicide with protective action, insecticide. Repellent to birds and rodents. Used for control of fungi in fruits, vines, vegetables and ornamentals. Controls scab in apples and pears. Dec 157°; soluble in H_2O (10 mg/l), insoluble in organic solvents; LD_{50} (rat orl) > 5200 mg/kg. *Agrimont; Diachem; Bayer; Pennwalt Holland; Makhteshim Chemical Works Ltd; Rhône Poulenc, Visplant; Rohm & Haas.*

1665   Zolone

2310-17-0   7489   218-996-2

$C_{12}H_{15}ClNO_4PS_2$

S-[(6-Chloro-2-oxo-3(2H)-benzoxazoly)methyl] O,O-diethyl phosphorodithioate.

S-6-chloro-2,3-dihydro-2-oxobenzoxazol-3-yl-methyl phosphorodithioate; O,O-diethyl phosphorodithioate S-ester with 6-chloro-3-(mercapto-methyl)-2-benzoxazolinone; ENT 27163; 11974 RP; Azofene; Rubitox. Emulsifiable concentrate containing 350 g/l phosalone; organo-phosphorus insecticide and acaricide for control of insects and spider mites on fruit and vegetables. mp = 45-48°; soluble in H_2O (10 mg/l), soluble in organic solvents; LD_{50} (rat orl) = 120-175 mg/kg. *Hortichem Ltd; Rhône Poulenc Crop Protection Ltd; Voltas.*

Insecticide Products

1666   Afrisect

Insecticide formulation. *Mitchell Cotts Chemicals Ltd.*

1667   Coopercote

Insecticidal varnish for paper, board etc. *The Wellcome Foundation Ltd.*

1668   Damoil

Phytonomic oil.

Dormant and summer spray oil for use as a contact insecticide. *Draxel Chemical Company.*

1669   Doff

Range of insecticides and herbicides; for horticultural/household use. *Doff Portland Ltd.*

1670   EP-2

O,O-Diethyl phosphorochlorodithioate.

Mainly used in the production of organophosphorus insecticides. *A/S Cheminova.*

1671   Ferox-Celotex

A proprietary trade name for Celotex which has been treated to resist attack by fungi and termites.

1672   Isotox

Insecticide seed treater. *Monsanto (Solaris).*

1673   Keriguards

Pellets containing fertilizer and insecticide in combination. *ICI Chem & Polymers Ltd.*

1674   Kil

Insecticides. *Fisons plc, Horticultural Div.*

1675   Killgerm® ULV 400

Pyrethin.

Contact insecticide. *Killgerm Chemical Ltd.*

1676 Killgerm® ULV 500
Mixture containing phenothrin and tetramethrin; for control of flying insects in agricultural premises. *Killgerm Chemicals Ltd.*

1677 Kival
Insecticide. *May & Baker Ltd.*

1678 Knave
Granular mixture of disulfoton and quinalphos; an organophosphorus insecticide. *Hortichem Ltd.*

1679 Larvex
A solution of sodium fluosilicate; a proprietary clothes-moth remedy.

1680 Lausofan
A hexamethylene ketone; used for destroying insects of the vermin type.

1681 Lizetan Spray
Insecticidal spray; used for control of biting and sucking pests. *Bayer AG.*

1682 Louse Powder
Insecticide. *Schering Agrochemicals Ltd.*

1683 Squadron
A suspension concentrate containing 100 g carbendazim [83601-81-4] and 275 g maneb [12427-38-8] per liter; systemic fungicide for cereals. *Quadrangle Agrochemicals.*

1684 Tanalith
Copper/chrome/arsenate waterborne wood preservative to prevent fungal decay and insect attack; for pressure treated timber for construction, fencing, agriculture and any application where timber requires protection. *Hickson & Welch Ltd.*

1685 Terminate
Bacillus thuringiensis wettable powder; applied by spray to control larvae of *Lepidopteran* insects. *Westbridge Research Group.*

1686 Terra-Systam
Systemic organo-phosphorus insecticides. *Murphy Chemical Co Ltd.*

1687 Texaco BQ
An insecticide having a petroleum base; used for killing the boll weevil.

1688 Unicrop Leatherjacket Pellets
γ-HCH.
An organochlorine insecticide, contains lindane. *Universal Crop Protection Ltd. See* lindane.

1689 Via Rasa
The calcium salt of p-toluenechlorosulfonamide; an insecticide.

1690 Vitavax RS Flowable
Carboxin + γ-HCH + thiram; a fungicide and insecticide dressing for oilseed rape. *Uniroyal Chemical Ltd. See* carboxin, γ-HCH, thiram.

1691 Westoran®
A trade name for a cleaning agent for cotton; it contains emulsified hydrocarbons, and is also used as an insecticide.

1692 Worm Ender
A natural occurring biological insecticide used for the control of worms and caterpillars on vegetables, ornamentals, fruit trees, shade trees, etc.; may be used up to the day of harvest. *Lawn & Garden Products Inc.*

Molluscicides

1693 Calcium Arsenate
7778-44-1   1686   231-904-5
$As_2Ca_3O_8$
Calcium arsenate.
Tricalcium arsenate; Pencal; Calcars. Used as insecticide and molluscicide. Slightly soluble in H_2O and dilute acids; LD_{50} (rat orl) = 298 mg/kg. *Mechema Chemicals Ltd.*

1694 Clonitrilide
1420-04-8   6602   215-811-7
$C_{13}H_8Cl_2N_2O_4$
2',5-Dichloro-4'-nitrosalicylanilide compound with 2-aminoethanol (1:1).
Bayluscid®; niclosamide ethanolamine; Bayer 73; Bayer 25648; Bayluscide; Clonitrilide; M 73; SR 73; Mollutox; Niclosamide; Phenasal ethanolamine salt; Yomesan. A molluscicide. mp = 204°; LD_{50} (rat orl) = 500 mg/kg. *Bayer AG.*

1695 Metaldehyde
108-62-3   5983   202-945-6
$C_8H_{16}O_4$
2,4,6,8-Tetramethyl-1,3,5,7-tetraoxacyclooctane.
metacetaldehyde; Ariotox; Antimilace; Acetaldehyde; Corry's slug death; Farmon Mini Slug Pellets; Gastratox 6G Slug Pellets; Halizan; Helarion;; Metacetaldehyde; Metason; META; MifaSlug; Namekil; Slug-tox; tetramer;; Ortho Metaldehyde 4% Bait; PBI Slug Pellets; Slug-Tox; UN1332; R-2,C-4,C-6,C-8-tetramethyl-1,3,5,7-tetroxocane; 2,4,6,8-tetramethyl-1,3,5,7-tetroxocane; Tetramethyl-1,3,5,7-tetroxocane; Totroxocane, 2,4,6,8-tetramethyl-. Molluscicide with contact and stomach action. Contact with the foot makes the mollusk torpid and induces an

increased secretion of mucus, leading to dehydration. Used for control of slugs and snails. The polymer has CAS RN 9002-91-9. Pellets containing 6% w/w metaldehyde; moluscicide with many crop uses; snail and slug bait. mp = 246°; sublimes 112-115°; soluble in H_2O (200 mg/l), C_6H_6, $CHCl_3$, insoluble in EtOH, Et_2O; LD_{50} (rat orl) = 630 mg/kg. *Truchem Ltd; Farm Protection Ltd; Fisons plc; Farmers Crop Chemicals Ltd.; Pan Britannica Industries Ltd.*

1696   Methiocarb
2032-65-7   6050   217-991-2
$C_{11}H_{15}NO_2S$
Phenyl-3,5-dimethyl-4-(methylthio)methylcarbamate.
Carbamic acid, methyl-, 4-(methylthio)-3,5-xylyl ester; B 37344; Bay 9026; Bayer 37344; BAY 37344; BAY 5024; BAY 9026; Club; Draza; DCR 736; Esurol; ENT 25,726; mercaptodimethur; Mesurol®; Mesurol phenol; methyl carbamic acid, 4-(methylthio)-3,5-xylyl ester; Metmercapturan; Metmercapturon; Methiocarbe. Molluscicide with neurotoxic action. Used for control of slugs and snails in a wide variety of agricultural situations. Pellets containing 4% w/w methiocarb; snail and slug bait. Formulated for variety of uses; especially as a molluscicide against slugs and snails as well as a seed dressing for repelling depredating birds; also as insecticide/acaricide against foliar-feeding caterpillars and sucking pests on various crops. mp = 121.5°; soluble in H_2O (27 mg/l), more soluble in organic solvents; LD_{50} (rat orl) = 20 mg/kg. *Bayer; Mobay; ICI AgroChemicals.*

1697   Mini Slugit Pellets
9002-91-9   5983   202-945-6
Metaldehyde.
Mota; Helarion; Quad Mini Slug Pellets; Tripart® Mini Slug Pellets. Molluscicide. A pelleted bait containing 6% w/w metaldehyde; used for control of slugs and snails. *Murphy Chemical Co Ltd.; Fisons plc; Quadrangle Agrochemicals; Tripart Farm Chemicals Ltd.; Schering Agrochemicals Ltd.*

Molluscicide Products

1698   Bug-Geta
Slug and snail pellets. *Monsanto (Solaris).*

1699   FBC Slug Destroyer
Molluscicide.

1700   Nobble
Aluminum sulfate + copper sulfate + sodium tetraborate.
Used for slug and snail control in field and glasshouse crops. *Fieldspray.*

Nematicides

1701   Aldicarb
116-06-3   223
$C_7H_{14}N_2O_2S$
2-Methyl-2-(methylthio)propanal O-[(methylamino)carbonyl]oxime.
aldicarbe; OMS 771; ENT 27093; Temik; UC 21149. Systemic insecticide, acaricide and nematicide with contact and stomach action. Cholinesterase inhibitor. mp = 99-100°; soluble in H_2O (0.6 g/100 ml), Me_2CO (35 g/100 ml), C_6H_6 (15 g/100 ml), xylene (5 g/100 ml), CH_2Cl_2 (30 g/100 ml); LD_{50} (frat orl) = 1 mg/kg. *Rhône Poulenc.*

1702   Aldoxycarb
1646-88-4
$C_7H_{14}N_2O_4S$
2-Methyl-2-methylsulfonylpropionaldehyde O-methylcarbamoyloxime.
aldoxycarbe; aldicarb sulfone; ENT 29261; sulfocarb; Standak; UC 21865. mp = 140-142°; soluble in H_2O (1 g/100 ml), $CHCl_3$ (3.2 g/100 ml), CH_3CN (7.4 g/100 ml), MeOH (3.0 g/100 ml), CH_2Cl_2 (4.1 g/100 ml), Me_2CO (5.0 g/100 ml). *Rhône Poulenc.*

1703   Cadusafos
95465-99-9
$C_{10}H_{23}O_2PS_2$
O-Ethyl S,S-bis(1-methylpropyl) phosphorodithioate.
Apache; FMC 67825; Rugby; Taredan. Used for control of nematodes and larvae of *inAgriotes spp. and *inPhthorimaea operculella in bananas, maize, potatoes, citrus, groundnuts, sugar cane; tobacco and vegetables. $bp_{0.8}$ = 112-114°; soluble in H_2O (0.0248 g/100 ml), freely soluble in organic solvents; LD_{50} (rat orl) = 37.1 mg/kg, (mus orl) = 71.4 mg/kg. *FMC.*

1704   Carbofuran
1563-66-2
$C_{12}H_{15}NO_3$
2,3-Dihydro-2,2-dimethyl-7-benzofuranyl methylcarbamate.
Valtox; Bay 70143; OMS 864; ENT 27164; Carbodan; Carbosip; Chinufur; Curaterr; FMC 10242; Furacarb; Furadan; Kenofuran; Nex; Pillarfuran; Rampart; Yaltox. Granular systemic carbamate insecticide and nematode containing 5% w/w carbofuran; used to control a wide range of soil and seedling pests including cabbot root fly, cabbage stem weevil, flea beetle, early aphids in brassicas, turnip root fly, fruit fly, millipedes, symphilids, beet leaf miner, springtails, wireworms, free-living nematodes, potato cyst

eelworms, carrot fly and carrot willow aphid. mp = 153-154°; sg = 1.18; soluble in H_2O (0.032 g/100 ml), Me_2CO (15 g/100 ml), CH_3CN (14 g/100 ml), CH_2Cl_2 (12 g/100 mol), cyclohexanone (9 g/100 ml), C_6H_6 (4 g/100 ml), EtOH (4 g/100 ml), DMSO (25 g/100 ml), DMF (27 g/100 ml); LD_{50} (rat orl) = 82 - 14.1 mg/kg, (dog orl) = 19 mg/kg. *Bayer plc; Makhteshim-Agan; Sipcam; Chemolimpex; FMC; All India Medical; KenoGard; Tripart; Pillar.*

1705 Chloropicrin
76-06-2
CCl_3NO_2
Nitrochloroform.
trichloronitromethane; Acquinite; Chlor-O-Pic; Dojyopicrin; Doroclor; Pic-Clor; Picrin-80; Tri-Chlor. Used as a soil disinfectant for control of nematodes; soil insects, soil fungi and weed seeds. Also used for fumigation of stored grain to control insects and rodents, and for glasshouse and mushroom house fumigation. Often used in combination with methyl bromide and other fumigants. mp = -64°; bp_{757} = 112.4°; soluble in H_2O (0.23 g/100 ml at 0°, 0.16 g/100 at 25°), miscible with most organic solvents. *Great Lakes; Mitsui Toatsu; Niklor.*

1706 Cloethocarb
51487-69-5
$C_{11}H_{14}ClNO_4$
2-(2-Chloro-1-methoxyethoxy)phenyl methyl carbamate.
BAS 2631; Lance. Used for control of corn rootworms on maize; Colorado beetles on potatoes and a variety of aphids, nematodes, caterpillars, scales etc. on groundnuts, potatoes, rice; sugar cane, tobacco, sorghum, coffee, wheat, oilseed rape and lucerne. Applied as a granular soil treatment, a seed treatment or bait pellets. mp = 80°; soluble in H_2O (0.13 g/100 ml), EtOH (15.3 g/100 ml), Me_2CO (> 100 g/100 ml), $CHCl_2$ (> 100 g/100 ml); LD_{50} (rat orl) = 35.4 mg/kg, (mus orl) 70.4 mg/kg, (rat sc) = 3000 mg/kg. *BASF.*

1707 Dazomet
533-74-4
$C_5H_{10}N_2S_2$
Tetrahydro-3,5-dimethyl-2H-1,3,5-thiadiazine-2-thione.
tiazon; DMTT; Basamid; Boszamet; Crag Fungicide 974; Dazoberg; Fongosan; N-521. Soil fumigant; acts by release of methyl isothiocyanate. Used as a soil sterilant, applied prior to planting

crops. Controls soil fungi, nematodes, germinating weed seeds and soil insects. Also used as a slimicide in paper mills. mp = 104-105°; soluble in H_2O (0.3 g/100 ml), cyclohexane (40 g/100 ml), $CHCl_3$ (39.1 g/100 ml), CH_3CO (17.3 g/100 ml), C_6H_6 (5.1 g/100 ml), EtOH (1.5 g/100 ml), Et_2O (0.6 g/100 ml); LD_{50} (rat orl) = 520 mg/kg, (mmus orl) = 455 mg/kg, (fmus orl) = 710 mg/kg. *BASF; Bos; Rhône-Poulenc; Diachem.*

1708 DCIP
108-60-1
$C_6H_{12}Cl_2O$
Bis(1-chloro-2-propyl) ether.
IK-141; Nemamort. Used for control of nematodes in vegetables, fruit, tobacco, ornamentals, tea and mulberries. bp = 187°; soluble in H_2O (0.012 g/100 ml); LD_{50} (mrat orl) = 503 mg/kg, (mmus orl) = 296 mg/kg, (fmus orl) = 536 mg/kg. *SDS Biotech.*

1709 1,3-Dichloropropane
142-28-9
$C_3H_6Cl_2$
1,3-Dichloropropane.
Trimethylene dichloride. Fumigant to control nematodes.

1710 Ethoprophos
13194-48-4
$C_8H_{19}O_2PS_2$
O-Ethyl S,S-dipropyl phosphorodithioate.
Mocap; VC9-104. Cholinesterase inhibitor. Non-systemic nematicidesand soil insecticide with contact action. $bp_{0.2}$ = 86-91°; sg = 1.094; soluble in H_2O (0.07 g/100 ml), very soluble (> 30 g/100 ml) in organic solvents; LD_{50} (rat orl) = 62 mg/kg, (rbt orl) = 55 mg/kg. *Rhône Poulenc; Mobil.*

1711 Fenamiphos
22224-92-6
$C_{13}H_{22}NO_3PS$
Ethyl 4-(methylthio)-m-tolyl isopropylphosphoramidate.
phenamiphos; Bay 68138; Nemacur. Cholinesterase inhibitor. Systemic nematicide with contact action.Absorbed by the roots and translocated to the leaves. Used for control of ecto- and endoparasitic, free-living, cyst-forming and root-knotnematodes. mp = 49.2°; sg = 1.15; soluble in H_2O (0.07 g/100 ml), readily soluble in CH_2Cl_2, iPrOH, C_7H_8, sparingly soluble in C_6H_{14}; LD_{50} (mrat orl) = 15.3 mg/kg, (frat orl) = 19.4 mg/kg, mus orl) = 22.7 mg/kg, (dog,cat orl) 10 mg/kg. *Bayer; Mobay.*

1712   Fensulfothion

115-90-2

$C_{11}H_{17}O_4PS_2$

O,O-Diethyl O-[p-(methylsulfinyl)phenyl]
phosphorothioate.

Dasanit; Bay 25141; S767; Dasanit(R); Dansanit;
Fonsulfothion; Terracur P; Daconit; Agricur;
Chemagro 25141; DMSP; OMS 37; ENT 24945.
Nematode and insecticide with primarily contact
action; some systemic activity. Soil application to
control nematodes. $bp_{0.01}$ = 138-141°; sg = 1.202;
soluble in H_2O (0.15 g/100 ml), soluble in most
organic solvents; LD_{50} (mrat orl) = 10.5 mg/kg, (frat
orl) = 2.2 mg/kg, (gpg orl) = 9 mg/kg. Bayer.

1713   Fosthiazate

98886-44-3

$C_9H_{18}NO_3PS_2$

O-Ethyl S-(1-methylpropyl) (2-oxo-3-thiazolidinyl)-
phosphonothioate.

IKI 1145; fosthiazate. Cholinesterase inhibitor,
used for control of nematodes. Ishihara Sangkyo.

1714   Isazofos

42509-80-8

$C_9H_{17}ClN_3O_3PS$

Diethyl O-(5-chloro-1-(1-methylethyl)-1H-1,2,4-
triazol-3-yl) phosphorothioate.

Miral; isazophos; Brace; Triumph; CGA-12223;
Triumph 4E; Victor. Cholinesterase inhibitor used
for control of root nematodes in bananas, citrus,
cotton, beet, rice, maize, sugar and vegetables.
Applied to the soil, also controls soil-dwelling
insects. $bp_{0.001}$ = 100°; soluble in H_2O (0.025
g/100 ml), miscible with most organic solvents;
LD_{50} (rat orl) = 40-60 mg/kg, (mrat sc) > 3100
mg/kg, (frat sc) = 118 mg/kg. Ciba-Geigy.

1715   Metam Sodium

137-42-8

$C_2H_4NNaS_2$

Methylcarbamodithioic acid sodium salt.

Vapam; metham; A7 Vapam; Busan 1020;
Karbation; Maposol; Metam-Fluid BASF; Nemasol;
Solasan 500; Sometam; Trimaton; VPM; Metam
sodium; carbam; SMDC; Metam S.A.U.; Roo-Pru;
Gatorooter; Vaporooter; Basamid-fluid; Carbam,
sodium salt; N-869; Sistan; Woodfume vapam;
Arapam; Limatox; Monam. Soil fumigant. Acts by
decomposition to methyl isothiocyanate. A soil
sterilant, applied prior to planting of edible crops;
for use in control of nematodes, weed seeds, soil
fungi, and soil insects. Soluble in H_2O (72.2 g/100
ml); moderately soluble in MeOH, EtOH;
insoluble in most other organic solvents; LD_{50}
(mrat orl) = 1800 mg/kg, (frat orl) = 1700 mg/kg,
(mus orl) = 285 mg/kg, (rbt sc) = 1300 mg/kg.

Aragonesas; BASF; Buckman; Diachem; ICI;
Pennwalt Holland; Rhône-Poulenc; United Agri
Products; Universal Crop Protection; Visplant.

1716   Methyl Isothiocyanate

556-61-6

C_2H_3NS

Isothiocyanatomethane.

Methyl mustard oil; Trapex; Degussa methyl
isothiocyanate; MTC; Biomet 33; MENCS. Multi-
purpose soil fumigant used for control of
nematodes, soil fungi, soil insects and weed seeds.
mp = 35-36°; bp = 118-119°; soluble in H_2O (0.82
g/100 ml), readily soluble in all common organic
solvents; LD_{50} (rat orl) = 175 mg/kg, (mmus orl) =
90 mg/kg, (rat sc) = 2780 mg/kg, (mmus sc) =
1870 mg/kg, (rbt sc) = 263 mg/kg. Schering.

1717   Oxamyl

23135-22-0

$C_7H_{13}N_3O_3S$

N',N'-Dimethyl-N-[(methylcarbamoyl)oxy]-1-
thiooxamimidic acid methyl ester.

Vydate; Oxamyl; Thioxamyl; DPX 1410; Vydate L;
dioxamyl. Contact and systemic insecticide,
acaricide and nematicide. Absorbed by roots and
foliage with translocation. Used for control of
nematodes in ornamentals, fruit trees, vegetables,
curcubits, beet, bananas, pineapples, groundnuts,
cotton, soybeans, tobacco, potatoes and other
crops. mp = 108-110°; sg = 0.97; soluble in H_2O
(28 g/100 ml), MeOH (144 g/100 ml), EtOH (33
g/100 ml), iPrOH (11 g/100 ml), Me_2CO (67 g/100
ml), C_7H_8 (1 g/100 ml); LD_{50} (rat orl) = 5.4 mg/kg,
(rbt sc) = 2960 mg/kg. Du Pont.

1718   Phorate

298-02-2

$C_7H_{17}O_2PS_3$

O,O-Diethyl S-(ethylthio)methyl
phosphorodithioate.

Thimet; Thimet(R); American Cyanamid 3911; EI
3911; CL 35,024; Rampart. Systemic insecticide
and acaricide with contact and stomach action.
Has some nematicidal activity. Cholinesterase
inhibitor. Used for control of some nematodes in
brassicas, beetroot, sugar beet; fooder beet,
carrots, field beans; broad beans, celery, maize,
sorghum, potatoes, tomatoes, hops, soybeans,
sunflowers, sugar cane, lucerne, cotton, coffee,
rice, groundnuts and some ornamentals. mp < -
15°; $bp_{0.8}$ = 118-120°; soluble in H_2O (0.005 g/100
ml), most organic solvents; LD_{50} (mrat orl) = 3.7
mg/kg, (frat orl) = 16 mg/kg, (mrat sc) = 6.2 mg/kg,
(frat sc) = 2.5 mg/kg, (gpg dsc) = 20-30 mg/kg.
Am. Cyanamid; MTM; Pesticides India; Voltas.

1719   Terbufos

13071-79-9

$C_9H_{21}O_2PS_3$

S-tert-Butylthiomethyl O,O-diethyl phosphorodithioate.

Counter 15G; AC 92100; Counter; Contraven; ST 100. Cholinesterase inhibitor; soil insecticide and nematicide with contact and stomach action. For control of nematodes in beet, maize, cotton, vines, peppers, soybeans, sorghum, onions, potatoes, cabbages, ornamentals, tobacco and bananas. mp = -29.2; $bp_{0.01}$ = 69°; sg = 1.105; poorly soluble in H_2O (0.00045 g/100 ml), readily soluble in most organic solvents; LD_{50} (rat orl) = 1.6 mg/kg, (mus orl) = 5.0 mg/kg, (rat sc) = 9.8 mg/kg, (rbt sc) = 1.0 mg/kg. *Am. Cyanamid.*

1720   Triazophos

24017-47-8

$C_{12}H_{16}N_3O_3PS$

O,O-Diethyl O-(1-phenyl-1H-1,2,4-triazol-3-yl) phosphorothioate.

Triazophos; Hostathion. Cholinesterase inhibitor. Broad-spectrum insecticide and caricide with contact and stomach action. Possesses some nematicidal activity. Used to control some nematodes in ornamentals and strawberries and as a bulb dip for tulips and garlic. mp = 182°; soluble in H_2O (0.003 g/100 ml), C_6H_{12} (> 0.1 g/100 ml), CH_2Cl_2 (5-10 g/100 ml), iPrOH (0.2-0.5 g/100 ml), C_7H_8 (2-5 g/100 ml); LD_{50} (rat orl) = 100-200 mg/kg, (rbt sc) > 5000 mg/kg. *Hoechst.*

Plant Growth Regulators

1721   Alar

1596-84-5   2874   216-485-9

$C_6H_{12}N_2O_3$

Butanedioic acid mono(2,2-dimethylhydrazide).

Daminozide; B-Nine; Dazide; DMASA; Alar; Aminozide; B-NINE; SADH; B 995; Dimas; Alar-85; Kylar; dimethylaminosuccinamic acid; NINE. Water soluble powder containing 85% daminozide, a plant growth regulator. Absorbed by leaves with translocation throughout the plant. Used on apples to restrict vegetative growth. mp = 157-164°; soluble in H_2O (100 g/l), less soluble in organic solvents; LD_{50} (rat orl) = 8400 mg/kg. *Murphy Chemical Co Ltd.; Fargro Ltd.; Fine Agrochemicals Ltd. See daminozide.*

1722   Betapal Concentrate

120-23-0   6484   204-380-0

$C_{12}H_{10}O_3$

(2-Naphthyloxy) acetic acid.

(2-naphthyloxy)acetic acid; BNOA; Naphthoxy-acetic acid; Phyomone; (2-naphthyloxy)acetic acid; BNOA; Phyomone; (β-Naphthyloxy)acetic acid. Soluble concentrate containing 16 g/l (2-naphthyloxy)acetic acid; plant growth regulator. *Synchemicals Ltd. See (2-naphthyloxy)acetic acid.*

1723   Cerone

16672-87-0   3777   240-718-3

$C_2H_6ClO_3P$

2-Chloroethyl phosphonic acid.

Ethephon; Cerone; Bromeflor; Ethrel; Ethrel C; Ethrel E; Ethrel R; Florel; Prep; Arvest; (Chloroethyl)phosphonic acid; Flordimex; 2-Chloroethanephosphonic acid. Growth regulator containing 480 g 2-chlorethyl-phosphonic acid (Ethephon) per liter; used for winter barley. (Sold under license from Union Carbide) A plant growth regulator containing ethephon. *(Sold in UK on behalf of Amchem Prods. Inc.) ICI Chem & Polymers Ltd; Embetec Cropt Protection Ltd.; ICI Agrochemicals; A H Marks & Co Ltd.; Kommer-Brookwick Ltd. See ethephon.*

1724   Chlormequat Chloride

999-81-5   2153   213-666-4

$C_5H_{13}Cl_2N$

Chlormequat chloride.

Chlorocholine chloride; 2-Chloro-N,N,N-trimethylethanaminium chloride; (2-chloroethyl)-trimethylammonium chloride; choline dichloride; AC 38555; CCC; Cycocel; New 5C Cycocel; Cycogan; ABM Chlormequat 40, 72.5; Ashlade 4-60 CCC, 700 CCC; Atlas Chlormequat 46, 700; CCC 700; Cleanacres PDR 675; Clifton Chlormequat 46; Cycogan; Fargro Chlormequat; Farmacel, Farmacel 645; Headland Swift; Hyquat 70, 75; Chafer 5C Chlormequat; Mandops Barleyquat B; Mandops Bettaquat B; Mandops Spring Poquaternary; MSS Chlormequat 40, 46, 60, 70; Portman Chlormequat 400, 460, 600, 700; Power 64, 640, 700; Quadrangle Chlormequat 700; Standup; Star Chlormequat; Titan; Tripart® Brevis; Tripart® Chlormequat 460; Terpal® CC, M. Soluble concentrate containing 400 or 725 g/l chlormequat, a plant growth regulator. Improves resistance against lodging of oats, rye, and wheat. mp = 241° (dec); soluble in H_72O, EtOH; insoluble in organic solvents; LD_{50} (mus orl) = 54 mg/kg. *ABM Chemicals Ltd; Atlas Interlates Ltd.; Farmers Crop Chemicals Ltd.; BASF AG; Agan Chemical Manufacturers Ltd.; Cleanacres Ltd.; Clifton Chemicals Ltd.; Agan Chemical Manufacturers Ltd.; Fargro Ltd.; Farm Protection Ltd.; ICI Chem & Polymers Ltd.; WBC Technology Ltd., Agrichem (International) Ltd.; Mandops (UK) Ltd.; Mirfield Sales Services Ltd.; Portman Agrochemicals Ltd.; Kommer Brookwick Ltd.; Quadrangle Agrochemicals; L W Vass (Agricultural) Ltd.; Star Agrochem Ltd.; Schering Agrochemicals Ltd.; Tripart Farm Chemicals Ltd.; BASF AG; BritAg Industries Ltd.; Cyanamid of Great Britain Ltd.; Portman Agrochemicals Ltd.*

1725   Chlorpropham

101-21-3   2240   202-925-7

$C_{10}H_{12}ClNO_2$

Isopropyl m-chlorocarbanilate.

Warefog; Mirvale; Furloe; Chloro-IPC; Isopropyl N-(3-chlorophenyl) carbamate; Spud-Nic; Tater-pex; CIPC; Sprout Nip; Chloropropham; (3-chlorophenyl)carbamic acid 1-methylethyl ester; Chlorpropham CIPC; MSS CIPC; N-3-Chloro-phenylisopropylcarbamate; Furloe Chloro IPC 4EC; Furlow Chloro IPC 20G; Beet; Bud-Nip; Beet-Kleen; Chlorocarbanilic acid isopropyl ester; Elbanil; Fasco WY-HOE; Jack Wilson chloro 51 oil; Liro cipc; Nexoval; Preweed; Spud-nip; Stopgerme-S. Supplied as a foggable solution; for controlling sprouting in ware potatoes. Carbamate herbicide and sprout depressant for use on stored potatoes. *Ciba-Geigy Agrochemicals; Wheatley Chemical Co Ltd; Dean Agrochemicals Ltd; Mirfield Sales Services Ltd.*

1726   Chryzoplus, Chryzopon, Chryzosan, Chryzotek

133-32-4   4995   205-101-5

$C_{12}H_{13}NO_2$

4-Indol-3-ylbutyric acid.

4-(3-indole)butyric acid; indolebutyric acid; 3-indolebutyric acid; 1H-indole-3-butanoic acid; Hormodin; Seradix; hormodin; IBA. For promotion and acceleration of root formation in plant clippings. mp = 123-125°; insoluble in H_2O, soluble in organic solvents; LD_{50} (mus ip) = 100 mg/kg. *Fargro Ltd.; Biosynth AG; Penta Mfg.; Pfaltz & Bauer; Spectrum Chem. Mfg.; Embetec Crop Protection Ltd.* See indolebutyric acid.

1727   Clopyralid

702-17-6   2462   216-935-4

$C_6H_3Cl_2NO_2$

3,6-Dichloro-2-picolinic acid.

Hadranol; Lontrel; Stinger; Dowco 290; 3,6-di-chloro-2-pyridinecarboxylic acid; Lontril F; Lontril T; 3,6-dichloropyridine-carboxylic acid. Plant growth regulator. *Makhteshim Chemical Wrks Ltd.*

1728   Cutlass

18467-77-1   3243   242-348-8

$C_{12}H_{18}O_7$

Diprogulic acid.

Dikegulac; Atrinal. Contains dikegulac; growth regulator for hedges. Soluble concentrate of 200g dikegulac sodium per liter; plant growth regulator for hedges. *ICI Garden Products; Rhône-Poulenc Environmental Prods. Ltd.* See dikegulac.

1729   Gibberellic Acid

77-06-5   4426   201-001-0

$C_{19}H_{22}O_6$

2,4a,7-Trihydroxy-1-methyl-8-methylenegibb-3-ene-1,10-dicarboxylic acid 1,4a-lactone.

Gibberellic acid; Gibberellin A3; Activol; Gibberellin 1; 2β-hydroxygibberellin 1; Gibb-tabs; Grocel; Regulex; trihydroxy-1-methyl-8-methylenegibb-3-ene-1,10-dicarboxylic acid, 1,4a-lactone. A plant growth regulator; increases cropping in apples and pears. mp = 233-235°; $[\alpha]_D^{19}$ = 86° (c = 2.12); soluble in H_2O, Et_2O; more soluble in EtOH, Me_2CO. *ICI Chem & Polymers Ltd.*

1730   Isoxaben

82558-50-7   5256

$C_{18}H_{24}N_2O_4$

N-(3-(1-Ethyl-1-methylpropyl)-5-isoxazolyl)-2,6-dimethoxybenzamide.

Gallery; benzamizole; EL-107; Flexidor. Suspension concentrate containing 500 g isoxaben per liter. A plant growth regulator used for control of annual dicotyledons in cereals, grass and fruit. *DowElanco Ltd.*

1731   Maleic Hydrazide

123-33-1   5745   204-619-9

$C_4H_4N_2O_2$

6-Hydroxy-2H-pyridazin-3-one.

Chiltern Fazor; Malazide; Mazide 25; MSS MH18; Regulox K; Royal Slo-Gro; 1,2-dihydropyridazine-3,6-dione; 6-hydroxy-3(2H)-pyridazinone. A tree growth inhibitor containing 185 g/liter maleic hydrazide (as the potassium salt); used to control shoots on the trunk and suckers around the base of street trees; it also inhibits the development of buds on the trunk, which remain dormant following treatment. A plant growth regulator for grass and to reduce bud growth in trees, hedges and vegetables. *Rhône-Poulenc Environmental Prods. Ltd; Rhône-Poulenc Environmental Prods. Ltd.; Bos Chemicals Ltd; Chiltern Farm Chemicals Ltd; Fisons plc, Horticultural Div.; Synchemicals Ltd.; Mirfield Sales Services Ltd.; Uniroyal Chemical Ltd.; Rhône-Poulenc; Fair Products; Uniroyal; Synchemicals; Pennwalt Holland.*

1732   Mefluidide

53780-34-0   5846   258-767-4

$C_{11}H_{13}F_3N_2O_3S$

N-[2,4-Dimethyl-5-[[(trifluoromethyl)sulfonyl]-amino]phenyl]acetamide.

Echo; Embark; MBR12325; Mowchem; Trimcut. Soluble concentrate containing 240 g/l mefluidide; grass growth suppressant. Plant growth regulator and herbicide which inhibits growth and development of grasses. Used in lieu of grass cutting, for example, on road verges and embankments. mp = 183-185°; soluble in H_2O (180 mg/l), more soluble in organic solvents; LD_{50} (rat orl) >4000 mg/kg. *ICI Agrochemicals Professional Prods.*

1733 Mepiquat Chloride

24307-26-4   5903   246-147-6

$C_7H_{16}ClN$

1,1-Dimethylpiperidinium chloride.

Pix® ULV; Mepiquat; Pix; 1,1-dimethylpiperidinium chloride; Dimethyl-piperidinium chloride; PIX DF; N,N-Dimethyl-piperidinium chloride; BAS-083; BAS-85559X. Plant growth regulator for reduction of undesired vegetative growth of cotton, better boil retention, earlier maturity; improves yield and market quality of garlic and onions Plant growth regulator, used in combination with ethephon to control growth of cereal crops. mp > 350°; LD_{50} (rat orl) = 1420 mg/kg. *BASF AG; BASF plc; Clifton Chemicals Ltd.*

1734 Metsulfuron-methyl

74223-64-6   6244

$C_{14}H_{15}N_5O_6S$

Methyl 2-[[[[(4-methoxy-6-methyl-1,3,5-triazin-2-yl)amino]carbonyl]amino]sulfonyl]benzoate.

Escort; metasulfron methyl; metsulfuron methyl ester; DPX-T6376; Gropper; ALLIE; Ally; Metsulfuron-methyl; Ally®. Plant growth regulator, used for control of annual dicotyledons in cereals. *DuPont UK.*

1735 Naphthylacetic Acid

86-87-3   6458   201-705-8

1-Naphthalene acetic acid.

Rooting Powder; 1-Naphthylacetic acid; Planofix; alpha-Naphthylacetic Acid; Naphthaleneacetic acid; Fruitone; NAA; Naphthalene-1-acetic acid; 1-Naphthaleneacetic acid. Powder containing 1-naphthylacetic acid; rooting powder. *Vitax Ltd.*

1736 Paclobutrazol

76738-62-0   7118

$C_{15}H_{20}ClN_3O$

(R*,R*)-(±)-β-[(4-Chlorophenyl)methyl]-α=(1,1-dimethylethyl-1H-1,2,4-triazole-1-ethanol.

Bonzi; Cultar; Bounty; Clipper; Club; Drize; Holdfast; Inevitan; Molass; Oryze; Parlay; PP 333; Predict; Smarect; Trimmid; ICI-PP-333; PP-333. Suspension concentrate containing 4 g/l paclobutrazol; plant growth regulator for use on ornamental plants. Plant growth regulator, gibberellin biosynthesis inhibitor. Used on fruit trees, flowers, turf and rice. mp = 165°; d = 1.22; soluble in H_2O (35 mg/l), more soluble in organic solvents; LD_{50} (rat orl) = 2000 mg/kg. *ICI Chem & Polymers Ltd.*

1737 Quindoxin

2423-66-7   219-352-3

$C_8H_6N_2O_2$

Quinoxaline-1,4-dioxide.

A growth promoter.

1738 Rhizopon A, AA

87-51-4   4994   201-748-2

$C_{10}H_9NO_2$

1H-Indole-3-acetic acid.

Indol-3-yl acetic acid; Heteroauxin; 3-Indoleacetic Acid; β-Indoleacetic Acid; Indoleacetate; IAA; (Indol-3-yl)acetic acid; Indolyl-3-acetic acid; Indolylacetic acid; Rhizopin; Skatole carboxylic acid. A root growth promoter in either powder or tablet form. mp = 165-169°; poorly soluble in H_2O, $CHCl_3$; soluble in EtOH, Me_2CO, Et_2O. *Fargro Ltd.*

1739 Somon

3926-62-3   2162   223-498-3

$C_2H_2ClNaO_2$

Chloroacetic acid sodium salt.

Sodium chloroacetate; Monoxone; Sodium mono-chloroacetate; SMA; SMCA. Powder containing 96% w/w sodium monochloroacetate; annual dicotyledons control in various horticultural crops. *Hortichem Ltd.* See sodium monochloracetate.

1740 Tachigaren 70

10004-44-1   4905   233-000-6

$C_4H_5NO_2$

5-Methyl-3(2H)-isoxazolone.

Hymexazol; 3-hydroxy-5-methylisoxazole; F-319; RTY-319. Fungicide for pelleting sugar beet seed. mp = 84-85°; soluble in H_2O (8.5 g/100 ml), very soluble in organic solvents; LD_{50} (rat orl) = 4678 mg/kg. *Sumito Chemical (UK) plc.*

1741 Tecgran

117-18-0   204-178-2

$C_6HCl_4NO_2$

1,2,4,5-Tetrachloro-3-nitrobenzene.

Arena; Bygran; Easytec; Fusarex; Hickstor; Hystore; Hytec; Nebulin; Tubodust; Tubostore. Fungicide with protective and curative action. Tecgran is provided as granules or dispersible powder containing tecnazene; protectant fungicide and potato sprout suppressant. Tecnazene in liquid fogging solution; used for controlling sprouting and dry rot in stored potatoes. mp = 99°; bp = 304° (dec); soluble in H_2O (0.44 mg/l), more soluble in organic solvents; LD_{50} (rat orl) = 2047 mg/kg. *Atlas Interlates Ltd; ICI; Wheatley Chemical Co Ltd.; Dean Agrochemicals Ltd.*

1742 Thifensulfuron-methyl

79277-27-3

$C_{12}H_{13}N_5O_6S_2$

3-[[[[(4-Methoxy-6-methyl-1,3,5-triazin-2-yl)-amino]carbonyl]amino]sulfonyl]-2-thiophenecarboxylic acid methyl ester.

Harmony Extra; DPX-M6316; Harmony; INM-6316; Pinnacle; Thiameturon methyl ester;

Thiameturon-methyl;Thifensulfuron Me. Controls annual dicotyledons in cereals.

1743   Tre-Hold
2122-70-5   218-332-1
$C_{14}H_{14}O_2$
Ethyl-1-naphthyl acetate.
ethyl 2-(1-naphthyl) acetate; ethyl 1-naphthalene acetate; Tipoff. Plant growth regulator; inhibits sprouting at pruning points. bp_3 = 158-160°; d_{25} = 1.106; insol in H_2O, soluble in organic solvents; LD_{50} (rat orl) = 3850 mg/kg. AH Marks & Co Ltd.

1744   Tributylphosphorotrithioate
78-48-8   201-120-8
$C_{12}H_{27}OPS_3$
S,S,S-Tributyl phosphorotrithioate.
Tribufos;   Tributylphosphorotrithioate;Tribufos; S,S,S-Tributyltrithiophosphate; DEF®; De-Green; E-Z-Off D; Easy off-D; Butyl phosphorotrithioate; S,S,S-tributyl   phosphorotrithioate; Phosphorotrithioic acid S,S,S-tributyl ester; Deleaf defoliant; Folex 6EC; FOS-FALL "A"; Ortho phosphate defoliant; Tribuphos. Plant growth regulator which acts as a defoliant. Used on cotton to facilitate harvesting. mp < -25°; $bp_{0.3}$ = 150°; d^{20} = 1.057; n_D^{25} = 1.532; poorly soluble in H_2O (2.3 mg/l), more soluble in organic solvents; LD_{50} (rat orl) = 233 mg/kg. Bayer AG.

Plant Growth Regulators containing 1-Naphthylacetic Acid

1745   Synergol
Mixture of dichlorophen, 4-indol-3-yl butyric acid, and 1-naphthylacetic acid; used to promote rooting of cuttings. Hortichem Ltd. See dichlorophen, 4-indol-3-yl butyric acid, 1-naphthylacetic acid.

Plant Growth Regulators containing 2-Naphthyloxyacetic Acid

1746   Tomato Setting Spray
Aerosol spray containing 2-naphthyloxyacetic acid; setting spray for tomatoes. Vitax Ltd.

Plant Growth Regulators containing Chlormequat-Choline Chloride Mixtures

1747   Atlas 5C Chlormequat
Soluble concentrate containing 460 g chlormequat and 320 g choline chloride per liter; plant growth regulator. Atlas Interlates Ltd. See chlormequat, choline chloride.

1748   Chafer 5C Chlormequat
Soluble concentrate containing 640 g chlormequat and 64 g choline chloride per liter; plant growth

regulator for use in cereals and ornamentals. BritAg Industries Ltd. See chlormequat, choline chloride.

1749   Mandops Halloween, Hele Stone
Mixture of chlormequat and di-1-p-menthene. Plant growth regulator. Mandops (UK) Ltd. See chlormequat and di-1-p-menthene.

1750   New 5C Cycocel
A mixture of chlormequat chloride and choline chloride; a plant growth regulator for cereals, linseed, winter oilseed rape, ornamentals. BASF plc; BASF AG. See chlormequat chloride, choline chloride.

1751   New 5C Cycocel
Soluble concentrate containing 645 g chlormequat and 32 g choline chloride per liter; plant growth regulator for use in cereals and ornamentals. Cyanamid of Great Britain Ltd. See chlormequat chloride, choline chloride.

1752   Pacer
Soluble concentrate containing 360 g chlormequat and 180 g 2-chloroethylphosphonic (ethephon) acid per liter; plant growth regulator for use in winter wheat. Farm Protection Ltd. See chlormequat, ethephon.

1753   Pentagan
Plant growth regulator containing chlormequat and choline chloride. Agan Chemical Manufacturers Ltd. See chlormequat, choline chloride.

1754   Portman 5C Chormequat
Chlormequat + choline chloride; plant growth regulator for use in cereals and ornamentals. Portman Agrochemicals Ltd. See chlormequat + choline chloride.

1755   Portman Supaquat
Chlormequat + choline chloride; plant growth regulator for use in cereals and ornamentals. Portman Agrochemicals Ltd. See chlormequat, choline chloride.

1756   Tripart® Chlormequat 5C
Mixture of 40.8% (w/w) chlormequat chloride [999-81-5] 460 g/liter and choline chloride [67-48-1] 320 g/liter (28.3% w/w); plant growth regulator for cereals and ornamentals. Tripart Farm Chemicals Ltd. See chlormequat chloride, choline chloride.

1757   Upgrade
Soluble concentrate containing 360 g/l chlormequat and 180 g 2-chloroethylphosphoric acid per liter; plant growth regulator for winter

wheat. *Rhône-Poulenc Crop Protection Ltd. See* chlormequat, 2-chloroethylphosphoric acid.

Plant Growth Regulators containing Chlorpropham-Propham Mixtures

1758 Atlas Indigo
Mixture of chlorpropam and propham; plant growth regulator to suppress sprout growth in stored potatoes. *Atlas Interlates Ltd. See* chlorpropam, propham.

1759 Pommetrol M
Mixture of chlorpropham and propham; a plant growth regulator for potato sprout growth suppression. *Sam Fletcher Agricultural Specialists. See* chlorpropham, propham.

1760 Power Gro-Stop
Mixture of chlorpropham and propham; a plant growth regulator for potato sprout growth suppression. *Kommer-Brookwick Ltd. See* chlorpropham, propham.

Plant Growth Regulators containing Cytokinin-Gibberelic Acid Mixtures

1761 Foliar TRIGGRR
A liquid containing plant growth regulators (cytokinin and gibberellic acid) and trace minerals; used to increase crop yields and quality; applied to the foliage; used for a wide variety of crops including cotton, soybeans, wheat, fruits and vegetables. *Westbridge Research Group.*

1762 Soil TRIGGRR
A liquid containing cytokinin, a plant growth regulator; used to increase crop yields and quality; applied to the soil; used for a wide variety of crops including corn, peanuts, sorghum, soybeans, fruits, and vegetables. *Westbridge Research Group.*

Plant Growth Regulators containing Ethephon-Mepiquat Chloiride Mixtures

1763 Power Platoon
Soluble concentrate containing 155 g 2-chlorethylphosphonic acid (ethephon) and 305 g mepiquat chloride per liter; plant growth regulator for cereal crops. *Kommer-Brookwick Ltd.*

1764 Terpal®
Soluble concentrate containing 155g 2-chlorethylphosphonic acid and 305g mepiquat chloride per liter; plant growth regulator for use in cereals. *BASF plc; BASF AG; Clifton Chemicals Ltd.*

1765 Upgrade
Soluble concentrate containing 360 g/l chlormequat and 180 g 2-chloroethylphosphoric acid per liter; plant growth regulator for winter wheat. *Rhône-Poulenc Crop Protection Ltd. See* chlormequat, 2-chloroethylphosphoric acid.

Plant Growth Regulators containing Maleic Hydrazide

1766 Mazide Selective
Soluble concentrate of 6 g dicamba, 200 g maleic hydrazide and 75 g MCPA per liter. Plant growth regulator for grass verges. *Synchemicals Ltd. See* dicamba, maleic hydrazide, MCPA.

1767 Retard
Potassium salt of maleic hydrazide; growth retardant for trees, shrubs, ivy and grass. *Draxel Chemical Company. See* maleic hydrazide.

Plant Growth Regulators containing Metsulfuron-Methyl

1768 Harmony®
Metsulfuron-methyl + thifensulfuron-methyl. Mixture of metsulfuron-methyl and thifensulfuron-methyl; for control of annual dicotyledons in cereals. *Du Pont Uk. See* metsulfuron-methyl, thifensulfuron-methyl.

Plant Growth Regulators containing Paraquat-Diaquat Mixtures

1769 Weedol
Weedkiller containing paraquat [1910-42-5] and diquat [85-00-7] for gardeners. *ICI Garden Products. See* paraquat, diquat.

1770 Parable
Soluble concentrate of 100 g diquat and 100 g paraquat per liter; used for weed control in field crops. *ICI Agrochemicals. See* diquat, paraquat.

Plant Growth Regulators containing Tecnazene

1771 Tripart® Arena Plus
6% (w/w) tecnazene [117-18-0] and 2% carbendazim [83601-81-4]; protectant fungicide and sprout suppressant for stored potatoes. *Tripart Farm Chemicals Ltd. See* tecnazene, carbendazim.

1772 Tripart® Arena 6 +xg TBZ
Dustable powder containing 6% (w/w) tecnazene [117-18-0] and 1.8% thiabendazole [148-79-8]; a protectant fungicide and potato sprout suppressant. *Tripart Farm Chemicals Ltd.*

Plant Growth Regulators containing Unspecified Ingredients

1773   Atlas Adherbe®
Crop chemical enhancer/additive. *Allied Colloids Ltd.*

1774   Atlas Indigo
Plant growth regulator used to protect vegetable crops. *Allied Colloids Ltd.*

1775   Birgin®
Sprout suppressant for potatoes in storage. *Bayer AG.*

1776   Clipper
Tree growth regulator. *ICI Chem & Polymers Ltd.*

1777   DUK-880
Lenacil + phenmedipham.
Lenacil and phenmedipham; used for control of annual dicotyledons in sugar beet. *DuPont UK.*

1778   Fotosensin
A condensation product of phthalic acid and resorcinol containing small proportions of copper and iron, Small quantities increase the root and stem growth of plants.

1779   Lawnturf®
Artificial turf made from polyvinylidene chloride filament, etc. Asahi Chem. Industry.

1780   Mandops Podquaternary
Chlormequat and di-1-p-menthene. plant growth regulator. *Mandops (UK) Ltd.*

1781   Mow-It-Less
Grass seed mixture. *ICI Garden Products.*

1782   Regal Crown
Combination of growth stimulators which enhances plant growth by accelerating root growth. *Regal Chemical Company.*

1783   Trimmit
Grass growth regulator. ICI Chem & Polymers Ltd.

1784   Upgrade
Soluble concentrate containing 360 g/l chlormequat and 180 g 2-chloroethylphosphoric acid per liter; plant growth regulator for winter wheat. Rhone-Poulenc Crop Protection Ltd. *See* chlormequat, 2-chloroethylphosphoric acid.

Rodenticides

1785   Alphachloralose
15879-93-3
$C_8H_{11}Cl_3O_6$
1,2-O-(2,2,2-Trichloroethylidene)-α-D-glucofuranose.
Chloralose; α-D-glucochloralose; Anhydroglucochloral; Alfamat; Aphosal; Chloroalosane; Dulcidor; Glucochloral; Glucochloralose; (R)-1,2-O-(2,2,2-trichloroethylidene)glucofuranose; Murex; Somio; (R)-1,2-O-(2,2,2-trichloroethyl-idene)-α-D-glucofuranose. Rodenticide. *Rentokil Ltd.*

1786   Brodifacoum
56073-10-0   1400   259-980-5
$C_{21}H_{33}BrO_3$
3-[3-(4'-Bromo[1,1'-biphenyl]-4-yl)-1,2,3,4-tetrahydro-1-naphthalenyl]-4-hydroxy-2H-1-benzopyran-2-one.
PP-581; WBA-8119; Ratak+; Talon Rat Bait; Havoc; Klerat; PP581; Ratak Plus; Talon; Volid. Rodenticide containing brodifacoum. Anticoagulant. mp = 228-230°; insoluble in H_2O, slightly soluble in organic solvents; LD_{50} (rat orl) = 270 µg/kg.

1787   Bromadiolone
28772-56-7   1403   249-205-9
$C_{30}H_{23}BrO_4$
3-[3-(4'-Bromo[1,1'-biphenyl]-4-yl)-3-hydroxy-1-phenylpropyl]-4-hydroxy-2H-1-benzopyran-2-one.
Bromone; Canadien 2000; Tamogam; LM-637; Super-caid; Super-rozol; Slaymor; Rentokil Deadline; Bromo-4-biphenylyl)ethyl)benzyl)-4-hydroxycoumarin; LM-637; Super-caid; 3-(3-(4'-Bromo-(1,1'-biphenyl)-4-yl)-3-hydroxy-1-phenyl-propyl)-4-hydroxycoumarin; Bromadiolone; Canadien 2000; Contrac; Maki; Ratimus; Tamogam; Boldo. Anticoagulant rodenticide. Ready-to-use, concentrated bait for control of rats, house mice. mp = 200-210°; λ_m = 260 nm ($E_{1\ cm}^{1\%}$ 538-582 EtOH); soluble in H_2O (19 mg/l), more soluble in organic solvents; LD_{50} (rat orl) = 1.125 mg/kg. *Ciba-Geigy Agrochemicals; Rentokil Ltd.*

1788   Cupric Arsenite
10290-12-7   2693   233-644-8
$CuHAsO_3$
Arsonic acid copper(2+) salt (1:1).
arsenious acid coppper(2+) salt(1:1); Scheele's green; pickle green. A commercial variety of Scheelés green (cupric arsenite). Used as a pigment, wood preservative, insecticide, fungicide, and rodenticide. Insoluble in H_2O, organic solvents.

1789 Difenacoum

56073-07-5   259-978-4

$C_{31}H_{24}O_3$

Biphenyl-4-yl-1,2,3,4-tetrahydro-1-naphthyl)-4-hydroxy-1(2H)-benzopyran-2-one.
WBA 8107; Ratak; Neosorexa PP580; Killgerm® Ratak Cut Wheat Rat Bait; Neosorexa; Ratak; Matikus; Talon. Anti-coagulant, used as a rodenticide. Difenacoum; a ready-to-use anticoagulant rodenticide. *Killgerm Chemicals Ltd.; Sorex Ltd.; ICI Garden Products; ICI Chem & Polymers Ltd.; ICI Agrochemicals; Lever Industrial Ltd.; Ace Chemicals Ltd.*

1790 Drat

3691-35-8   2204   223-003-0

$C_{23}H_{15}ClO_3$

Chlorophacinone.
2-[(4-Chlorophenyl)phenylacetyl]-1H-indene-1,3(2H)-dione; 2-[(p-chlorophenyl)phenylacetyl]-1,3-indandione; LM-91; Caid; Drat; Liphadione; Quick; Raviac; Rozol; Rozol; Karate. An oil formulation containing 2.5 g/liter chlorophacinone, a powerful anticoagulent rodenticide; controls black rats, brown rats, house mice, long-tailed field mice, voles and musk rats. mp = 140°. *Rhône-Poulenc Environmental Prods. Ltd.; Lever Industrial Ltd.*

1791 Killgerm®; Sewarin; P

81-81-2   10174   201-377-6

$C_{19}H_{16}O_4$

Warfarin.
4-Hydroxy-3-(3-oxo-1-phenylbutyl)-2H-1-benzo-pyran-2-one; 3-(α-acetonylbenzyl)-4-hydroxy-coumarin; 1-(4'-hydroxy-3'-coumarinyl)-1-phenyl-3-butanone; 3-α-phenyl-β-acetylethyl-4-hydroxy-coumarin; compound 42; WARF compound 42; Co-Rax; Rodex; Compound 42; Coumadin; D-con; 3-(alpha-acetonylbenzyl)-4-hydroxy-coumarin; Panwarfin; Sofarin; 4-hydroxy-3-(3-oxo-1-phenylbutyl)-2H-1-Benzopyran-2-one; Rodex; zoocoumarin; Co-Rax; Cov-R-Tox; Kypfarin; Liqua-Tox; RAX; Rodex-Blox; Tox-Hid; athrombink; arab rat death; brumolin; dethnel; eastern states duocide; fasco fascrat; kumader; kumadu; mar-frin; martin's mar-frin; maveran; 3-(alpha-phenyl-beta-acetylethyl)-4-hydroxycoumarin; 3-(1'-phenyl-2'-acetylethyl)-4-hydroxycoumarin; pro-thromadin; rat-b-gon; rat-gard; rat and mice bait; rat-mix; rat-o-cide 2; rat-ola; ratorex; rat-kill; Ratox; ratron; ratron g; rats-no-more; rat-trol; rattunal; rodafarin; ro-deth; Rosex; rough and ready mouse mix; solfarin; spray-trol brand rodentrol; temus w; twin ligh; vampirinip ii; vampirinip iii; Waran; w.a.r.f. 42; warfarat; warfarin plus; warf compound 42; warficide; Benzopyran-2-one, 4-hydroxy-3-(3-oxo-1-phenylbutyl)-; Marevan. 0.025% Warfarin on pin-head oatmeal, with sugar and mold inhibitor; rat and mouse killer. Warfarin;

bait for grey squirrels. mp = 161°; λ_m (water, pH 10) = 308 nm (ε 13610). *Killgerm Chemicals Ltd.; Mechema Chemicals Ltd.; The Wellcome Foundation Ltd.; Leo Fay Ltd.*

1792 Malladrite

16893-85-9   8769   240-934-8

Na_2SiF_2

Sodium fluosilicate.
sodium silicofluoride; sodium hexafluorosilicate; Sodium fluosilicate; Earwig bait; Silicate(2-), hexafluoro-, disodium; disodium hexafluorosilicate. Used in enamels, as a moth repellent, insecticide and rodenticide. *Alfa Aesar; Pfaltz & Bauer; ICN Biomedical Research Prods.*

1793 Phostoxin

20859-73-8   372   244-088-0

AlP

Aluminum phosphide.
Power Phosphine Pellets; Talunex. Reacts with water to give phosphine (PH_3). Used for gasing of rabbits and moles. Used for gasing of rabbits and moles. d^{15}_4 = 2.85; stable below 1000°. *Rentokil Ltd.; Kommer-Brookwick Ltd.*

1794 Photophor®

1305-99-3   1742   215-142-0

Ca_3P_2

Calcium phosphide.
Photophor; Calcium phosphide (Ca_3P_2); Polytanol; Calcium phosphide, Technical Grade. A trade name for a calcium phosphide, used for signal fires and as a rodenticide. mp = 1600°; d = 2.51; decomposed by H_2O. *Alfa Aesar; Pfaltz & Bauer; ICN Biomedical Research Prods.*

1795 Prussic Acid

74-90-8   4836   200-821-6

CHN

Hydrocyanic Acid.
prussic acid; formonitrile; Cyclon; HCN. Colorless or pale blue liquid or gas with a bitter almond odor detectable at 1 to 5 ppm. Present in apricot and peach pits in low concentrations, extremely toxic, used to exterminate rodents and insects. mp = -13°; bp = 26°; soluble in H_2O, LC_{50} (rat ihl 5 min) = 544 ppm.

1796 Racumin®

5836-29-3   227-424-0

$C_{19}H_{16}O_3$

4-Hydroxy-3-(1,2,3,4-tetrahydro-1-naphthalenyl)-2H-1-benzopyran-2-one.
Coumetralyl; 2H-1-Benzopyran-2-one, 4-hydroxy-3-(1,2,3,4-tetrahydro-1-naphthalenyl)-; Coumarin, 4-hydroxy-3-(1,2,3,4-tetrahydro-1-naphthyl)-; BAY 25634; Bay ene 11183 B; Bayer 25 634; Coumatetralyl; Cumatetralyl; Endox; Endrocid;

Endrocide; Ene 11183 B; 4-Hydroxy-3-(1,2,3,4-tetrahydro-1-naftyl)-cumarine; 4-Hydroxy-3-(1,2,3,4-tetrahydro-1-naphthalenyl)-2H-1-benzo-pyran-2-one; 4-Hydroxy-3-(1,2,3,4-tetra-hydro-1-naphthyl)coumarin; 4-Hydroxy-3-(1,2,3,4-tetra-hydro-1-naphthyl)cumarin; 3-(1,2,3,4-Tetrahydro-1-naphtyl)-4-hydroxy-coumarine; 3-(α-Tetralinyl)-4-hydroxycoumarin; 3-(α-Tetral)-4-oxycoumarin; 3-(D-Tetralyl)-4-hydroxycoumarin; 3-(α-Tetralyl)-4-hydroxy-coumarin; Rodentin; 3-(1,2,3,4-Tetra-hydro-1-naphthyl)-4-hydroxycumarin. Ready-to-use anticoagulant rodenticide. *Bayer AG, plc.*

1797   Ridene

3691-35-8   2204   223-003-0

$C_{31}H_{24}O_3$

2-[(4-Chlorophenyl)phenylacetyl]-1H-Indene-1,3(2H)-dione.

chlorophacinone; Topitox; Chlorophenyl)phenyl-acetyl)-1,3-indandione; Indandione, 2-((p-chlorophenyl)phenylacetyl)-; Indene-1,3(2H)-dione, 2-((4-chlorophenyl)phenylacetyl)-; Partox; 2-((p-Chlorophenyl)phenylacetyl)-1,3-indandione; LM-91; Caid; Drat; Liphadione; Quick; Raviac; Rozol; Sakarat Special; Skaterpax. An oil formulation containing 2.5% of chlorophacinone; an anticoagulant rodenticide; bait to control black rats, brown rats, house mice and voles. mp = 140°; λ_m = 325 nm (Me$_2$CO); poorly soluble in H$_2$O, soluble in organic solvents; LD$_{50}$ (rat orl) = 20 mg/kg. *Killgerm Chemicals Ltd.*

1798   Sodium Fluoride

7681-49-4   8762   231-667-8

FNa

Hydrofluoric acid sodium salt.

Disodium difluoride; Floridine; Florocid; Villiaumite; NaF; sodium hydrofluoride; sodium monofluoride; trisodium trifluoride; alcoa sodium fluoride; antibulit; cavi-trol; chemifluor; Credo; duraphat; fda 0101; f1-tabs; flozenges; fluoral; fluorident; fluorigard; fluorineed; fluorinse; fluoritab; fluorocid; fluor-o-kote; fluorol; fluoros; Flura; flura-gel; flura-loz; flurcare; flursol; fungol b; Gel II; gelution; Gleem; iradicav; karidium; karigel; kari-rinse; lea-cov; lemoflur; luride; luride lozi-tabs; luride-sf; nafeen; nafpak; na frinse; nufluor; ossalin; Ossin; osteofluor; pediaflor; pedident;pennwhite; pergantene; phos-flur; point two; predent; rafluor; rescue squad; Roach salt; sodium fluoride cyclic dimer; So-flo; stay-flo; studafluor; super-dent; t-fluoride; thera-flur; thera-flur-n; zymafluor; Les-cav. Fluoridation of municipal water, degassing steel, wood preservative, insecticide, fungicide, rodenticide, chemical cleaning, electroplating, glass manufacture, vitreous enamels, preservative for adhesives, toothpastes, disinfectants, dental prophylaxis. mp = 993°; bp = 1704°; d = 2.78; soluble in H$_2$O (4 g/100 ml); insoluble in organic solvents; LD$_{50}$ (rat

orl) = 0.18 g/kg. *Cerac; EM Industries; General Chem.; Hoechst Celanese; Solvay GmbH; Whiting, Peter Ltd.*

Rodenticide Products

1799   Final Flip

Pelleted rodenticide, single feed type, indoor or outdoor use, place pack size; chronic toxicant as the active material. *Colonial Products Inc.*

1800   Hyperkil Bait

Caciferol.

A rodenticide. *Antec international Ltd.*

1801   Klerat

Brodifacoum.

Rodenticide. *ICI Chem & Polymers Ltd.*

1802   Mouse Killer

Rodenticide. *Murphy Chemical Co Ltd.*

1803   Muritan

Compound used to control mice. *Bayer AG.*

1804   Rinoxin

A rodenticide. *Gerhardt Pharmaceuticals.*

1805   Scheele's acid

A 4% solution of hydrocyanic acid, HCN.

1806   Sorexa CD

Cupric Arsenite.

Caciferol + difenacoum; used for the control of mice in farm buildings. *Sorex Ltd.*

Slimicides

1807   Daracide

Slimicide for use in the paper and pulp industry. Grace Dearbom Ltd.

1808   Didecyl Dimonium chloride

7173-51-5   230-525-2

$C_{22}H_{48}ClN$

N-Decyl-N,N-dimethyl-1-Decanaminium chloride.

Arquad 10; Bardac 22; Decanaminium, N-decyl-N,N-dimethyl-, chloride; didecyl dimethyl ammonium chloride; Quaternium 12; BTC® 99. Low foaming algicide and slimicide for swimming pool and industrial water treatment. *Stepan; Stepan Canada.*

1809   Drewbrom

Aqueous solution of bromide ion; precursor for production of biocide used as disinfectant, sanitizer, bactericide, slimicide, and algicide in recirculating cooling water systems, once-through cooling water, and waste water treatment systems. *Drew Ind. Div.*

1810 Mycocide

A slime control agent. *Great Lakes Europe.*

1811 N,N-Dimethyl-N-tetradecyl-benzenemethanaminium chloride

139-08-2   205-352-0

$C_{23}H_{42}ClN$

Myristalkonium chloride.

Zephiramine; C14 benzyl dimethyl ammonium chloride; C14 dimethyl benzyl ammonium chloride; Benzenemethaminium, N-tetradecyl-N,N-dimethyl, chloride; Benzyl dimethyl tetradecyl ammonium chloride; C14-alkylbenzyldimethylammonium chloride; Roccal MC-14; BTC® 2565; Tetradecyl dimethyl benzyl ammonium chloride; Miristalkonium chloride; TDBAC. Algicide and slimicide for swimming pool and industrial water treatment. *Stepan; Stepan Canada.*

1812 Phenylmercuric Acetate

62-38-4   200-532-5

$C_8H_8HgO_2$

Phenylmercury acetate.

(acetato)phenylmercury; acetoxyphenylmercury; phenylmercury acetate; PMA; (acetato-O)phenylmercury; Phenylmercuric acetate; Phenylmercury

(II) Acetate; (Aceto)-phenylmercury; PMAC; PMAS; Gallotox; Liqui-phene; Phix; Mersolite; Tag HL-331; Nylmerate; Scutl; Riogen; Advacide PMA 18; Cosan PMA; Mergal A25; Metasol 30; Nildew AC 30; Nuodex PMA 18; Agrosan; Cekusil; Celmer; Hong Nien; Pamisan; Seedtox; Shimmerex; Unisan; acetic acid, phenylmercury deriv.; (acetoxymercuri)benz-ene; agrosan gn 5; algimycin; antimucin wdr; (acetoxymercurio)-benzene; Bufen; ceresan universal; contra creme; dyanacide; Femma; FMA; fungitox or; HL-331; hostaquik; kwiksan; leytosan; mersolite 8; norforms; phenmad; phenomercuric acetate; phenylmercuriacetate; pmacetate; PMAL; purasan-sc-10; puraturf 10; quicksan; quicksan 20; sanitized spg; SC-110; spor-kil; TAG; tag 331; trigosan; ziarnik; Anticon; Fungicide R; Fungitox; Meracen; Mercuron; Neantina. Metallo-organic compound; fungicide, herbicide, mildewcide for paints; slimicide in paper mills. *Allchem Industries; Atomergic Chemetals; W.A. Cleary; EM Industries.*

1813 Vancide® 51

Sodium dimethyldithiocarbamate (27.6%) and sodium 2-mercaptobenzothiazole (2.4%); fungicide for use as a preservative in latex and starch paste; bactericide for sol. cutting fluids and coolants; paper mill slimicide; Used in petrol. storage tanks, recirculating cooling towers, paper and paperboard, and cotton fabric. *R. T. Vanderbilt Co Inc.*

PART II

INDEXES

1 CAS Registry Number Index

CAS Registry Number Index

CAS RN	Name/Synonym	Record No.	CAS RN	Name/Synonym	Record No.
54-11-5	XL-All Insecticide	1663	108-60-1	DCIP	1708
56-38-2	Parathion	23	108-62-3	Metaldehyde	1695
57-09-0	Cetrimide	815	110-44-1	Sorbistat	480
57-13-6	Bubber Shet	43	114-26-1	Propoxur	1647
57-13-6	Urea	206	115-29-7	Thiodan	32, 1658
58-89-9	Lorexane	1644	115-32-2	Dicofol	9
60-51-5	Dimethoate	11	115-90-2	Terracur® P	1655
61-82-5	Aminotriazole Bayer	798	115-90-2	Fensulfothion	1712
62-38-4	Acticide PMA 100	376	116-06-3	Aldicarb	1701
62-38-4	Merpectogel	875	116-29-0	Tedion V-18	31
62-38-4	Phenylmercuric acetate	1812	117-18-0	Ashlade TCNB	383
62-73-7	Divipan	12	117-18-0	Tecgran	1741
62-73-7	Vapona	1660	120-12-7	Sterilite Hop Defoliant	58
63-25-2	Carbaryl	1630	120-23-0	Betapal Concentrate	1722
63-42-3	milk sugar	171	120-36-5	Dichloroprop	833
67-48-1	choline chloride	168	121-75-5	Malathion	16
74-90-8	Prussic Acid	1795	122-34-9	Simazine	910
75-99-0	Dalapon	827	122-42-9	Propham	902
76-03-9	Trichloroacetic acid	924	123-33-1	Maleic hydrazide	1731
76-06-2	Chloropicrin	1705	123-88-6	Atiran	384
76-87-9	Triphenyltin hydroxide	493	126-06-7	Halobrom	50
77-06-5	Gibberellic acid	1729	127-20-8	Dowpon	843
78-48-8	Tributylphosphorotrithioate	1744	128-04-1	Sodium Dimethyldithio-carbamate	478
81-81-2	Killgerm® Sewarin P	1791			
82-68-8	Quintozene	476	133-06-2	Vancide® 89	495
85-00-7	Diquat dibromide	841	133-07-3	Folpet	48
86-87-3	Naphthylacetic acid	1735	133-07-3	Folpan	435
87-51-4	Rhizopon A, AA	1738	133-32-4	Chryzoplus	1726
88-85-7	Tubotox	494	133-90-4	Naptol	887
88-85-7	Dinoseb	839	137-26-8	Thiram	488
92-52-4	Diphenyl	421	137-42-8	Metam Sodium	1715
93-65-2	Mecoprop-P	871	139-08-2	N,N-Dimethyl-N-tetradecyl-benzenemethanaminium chloride	1811
93-72-1	Silvex	909			
93-76-5	2,4,5-T	914			
94-13-3	Propylparaben	475	139-40-2	Propazine	901
94-26-8	Butylparaben	388	140-56-7	Fenaminosulf	428
94-74-6	MCPA	868	142-28-9	1,3-Dichloropropane	1709
94-75-7	2,4-D	824	142-59-6	Campbell's Nabam Soil Fungicide	394
94-81-5	MCPB	450, 869			
94-82-6	2,4-DB	825	144-21-8	Methar 30	878
94-96-2	ethyl hexanediol	47	144-55-8	Tronacarb Sodium Bicarbonate	172
97-23-4	Dichlorophen	417			
99-30-9	Dicloran	418	148-79-8	Thiabendazole	485, 1657
100-02-7	P.N.P	465	149-30-4	Thiotax	487
101-05-3	Anilazine	380	151-38-2	Panogen M	466
101-10-0	Bidisin	808	156-62-7	Lime nitrogen	55
101-21-3	Mirvale	453	288-32-4	Imidazole	1638
101-21-3	Chlorpropam	819	298-00-0	Parathion-methyl	24
101-21-3	Chlorpropham	1725	298-02-2	Terrathion	1656
103-34-4	Sulfasan	482	298-02-2	Phorate	1718
106-46-7	Kaydox	1641	299-84-3	Korlan	443, 1643

CAS Registry Number Index

CAS	Name	Page	CAS	Name	Page
314-40-9	Bromacil	811	1563-66-2	Carbofuran	1704
330-54-1	Diuron	842	1582-09-8	Agriphlan 24	793
330-55-2	Linuron	864	1596-84-5	Alar	1721
333-41-5	Diazinon Liquid	1636	1646-88-4	Aldoxycarb	1702
470-90-6	Birlane	40	1689-83-4	Ioxynil	865
470-90-6	Chlorfenvinphos	1631	1689-84-5	Bromoxynil	812
471-34-1	Mild Lime	57	1698-60-8	Chloridazon	816
485-31-4	Morocide	454	1702-17-6	Clopyralid	44, 821, 1727
520-45-6	Dehydroacetic acid	414	1715-40-8	Bromodan	42
532-32-1	Sodium Benzoate	477	1746-81-2	Aresin	800
533-74-4	Dazomet	1707	1762-95-4	Ammonium Thiocyanate	196, 799
534-52-1	Dekryll	828	1777-82-8	Myacide® SP	456
555-37-3	Kloben®	52	1861-32-1	Dacthal	826
555-37-3	Neburon	888	1861-40-1	Benfluralin	806
556-61-6	Methyl isothiocyanate	1716	1897-45-6	Bombardier	41
563-12-2	Ethion	1637	1897-45-6	Chlorothalonil	400
582-25-2	Potassium Benzoate	469	1910-42-5	Gramoxone	856
583-91-5	MHA	170	1912-24-9	Atrazine	803
590-00-1	Sorbistat K	481	1912-26-1	Trietazine	926
640-15-3	Thiometon	33	1918-00-9	Dicamba	831
709-98-8	Propanil	900	1918-02-1	Tordon	922
719-96-0	Fluorfolpet	430	1918-16-7	Propachlor	899
731-27-1	Euparen® M	427	1929-73-3	Planotox	896
759-94-4	EPTC	845	1982-47-4	Chloroxuron	818
834-12-8	Ametryn	797	2032-65-7	Mesurol®	17
886-50-0	Clarosan 1FG	820	2032-65-7	Methiocarb	1696
919-86-8	Demeton-S-methyl	8	2122-70-5	Tre-Hold	1743
919-86-8	Vassgro DSM	1661	2163-80-6	MSMA	885
950-10-7	Mephosfolan	874, 1645	2164-08-1	Lenacil	863
950-37-8	Methidathion	19	2164-17-2	Fluometuron	852
957-51-7	Diphenamid	840	2227-17-0	Dienochlor	10
999-21-3	Diallyl Maleate	1635	2303-16-4	Diallate	830
999-81-5	Chlormequat chloride	1724	2303-17-5	Triallate	923
1014-69-3	Desmetryn	829	2307-68-8	Pentanochlor	894
1071-83-6	Glyphosate	855	2310-17-0	Phosalone	25
1085-98-9	Dichlofluanid	416	2310-17-0	Zolone	1665
1113-02-6	Omethoate	22, 1646	2423-66-7	Quindoxin	1737
1194-65-6	Dichlobenil	832	2425-06-1	Captafol	395
1305-78-8	Calcium Oxide	390, 1629	2439-01-2	Quinomethionate	28
1305-99-3	Photophor®	1794	2439-10-3	Dodine FL, WP	423
1314-13-2	Vita Zinc	173	2593-15-9	Aaterra WP	375
1314-56-3	Superphosphate	205	2675-77-6	Chloroneb 65W Fungicide	399
1317-39-1	Cuprous Oxide	412	2921-88-2	Talon	1652
1317-65-3	Garden Lime	49	2939-80-2	Difolatan	419
1318-72-5	Kainite	200	3060-89-7	Metobromuron	880
1327-53-3	arsenic trioxide	382, 801, 1627	3337-71-1	Asulam	802
1330-43-4	Sodium Borate	203	3347-22-6	Delan-Col	415
1332-14-5	Cupric Sulfate, Basic	411	3383-96-8	Temephos	918
1332-40-7	Copper Oxychloride	403	3691-35-8	Drat	1790
1332-58-7	Bilt-Cote®	39	3691-35-8	Ridene	1797
1338-02-9	Copper Naphthenate	402	3813-05-6	Benazolin	805
1344-81-6	Orthorix	461	3878-19-1	Fuberidazole	436
1420-04-8	Clonitrilide	1694	3926-62-3	Sodium Monochloracetate	913
1563-66-2	Rampart	1648	3926-62-3	Somon	1739

CAS Registry Number Index

2 EINECS Number Index

EINECS Number Index

EINECS	Name/Synonym	Record No.	EINECS	Name/Synonym	Record No.
200-193-3	XL-All Insecticide	1663	203-768-7	Sorbistat	480
200-271-7	Parathion	23	204-043-8	Propoxur	1647
200-315-5	Bubber Shet	43	204-079-4	Thiodan	32, 1658
200-315-5	Urea	206	204-082-0	Dicofol	9
200-401-2	Lorexane	1644	204-114-3	Terracur® P	1655
200-480-3	Dimethoate	11	204-134-2	Tedion V-18	31
200-521-5	Aminotriazole Bayer	798	204-178-2	Ashlade TCNB	383
200-532-5	Acticide PMA 100	376	204-178-2	Tecgran	1741
200-532-5	Merpectogel	875	204-371-1	Sterilite Hop Defoliant	58
200-532-5	Phenylmercuric acetate	1812	204-380-0	Betapal Concentrate	1722
200-547-7	Divipan	12	204-390-5	Dichloroprop	833
200-547-7	Vapona	1660	204-497-7	Malathion	16
200-555-0	Carbaryl	1630	204-535-2	Simazine	910
200-559-2	milk sugar	171	204-542-0	Propham	902
200-655-4	choline chloride	168	204-619-9	Maleic hydrazide	1731
200-821-6	Prussic Acid	1795	204-659-7	Atiran	384
200-923-0	Dalapon	827	204-766-9	Halobrom	50
200-927-2	Trichloroacetic acid	924	204-828-5	Dowpon	843
200-990-6	Triphenyltin hydroxide	493	204-876-7	Sodium Dimethyldithio-	
201-001-0	Gibberellic acid	1729		carbamate	478
201-120-8	Tributylphosphorotrithioate	1744	205-087-0	Vancide® 89	495
201-377-6	Killgerm® Sewarin P	1791	205-088-6	Folpet	48
201-435-0	Quintozene	476	205-088-6	Folpan	435
201-579-4	Diquat dibromide	841	205-101-5	Chryzoplus	1726
201-705-8	Naphthylacetic acid	1735	205-123-5	Naptol	887
201-748-2	Rhizopon A, AA	1738	205-286-2	Thiram	488
201-861-7	Tubotox	494	205-352-0	N,N-Dimethyl-N-tetradecyl-	
201-861-7	Dinoseb	839		benzenemethanaminium	
202-163-5	Diphenyl	421		chloride	1811
202-264-4	Mecoprop-P	871	205-359-9	Propazine	901
202-271-2	Silvex	909	205-419-4	Fenaminosulf	428
202-273-3	2,4,5-T	914	205-547-0	Campbell's Nabam Soil	
202-307-7	Propylparaben	475		Fungicide	394
202-318-7	Butylparaben	388	205-620-7	Methar 30	878
202-360-6	MCPA	868	205-633-8	Tronacarb Sodium	
202-361-1	2,4-D	824		Bicarbonate	172
202-361-1	Verdone CDA	972, 1389	205-725-8	Thiabendazole	485, 1657
202-365-3	MCPB	450, 869	205-736-8	Thiotax	487
202-366-9	2,4-DB	825	205-790-2	Panogen M	466
202-377-9	ethyl hexanediol	47	205-861-8	Lime nitrogen	55
202-567-1	Dichlorophen	417	206-019-2	Imidazole	1638
202-746-4	Dicloran	418	206-050-1	Parathion-methyl	24
202-811-7	P.N.P	465	206-052-2	Terrathion	1656
202-910-5	Anilazine	380	206-082-6	Korlan	443
202-915-2	Bidisin	808	206-082-6	Korlan	1643
202-925-7	Mirvale	453	206-245-1	Bromacil	811
202-925-7	Chlorpropham	819, 1725	206-354-4	Diuron	842
202-945-6	Metaldehyde	1695	206-356-5	Linuron	864
202-945-6	Mini Slugit Pellets	1697	206-373-8	Diazinon Liquid	1636
203-103-0	Sulfasan	482	207-432-0	Birlane	40
203-400-5	Kaydox	1641	207-432-0	Chlorfenvinphos	1631

3 Name and Synonym Index

Name and Synonym Index

Name and Synonym Index

Name and Synonym Index

Name and Synonym Index

Name and Synonym Index

Name and Synonym Index

Name and Synonym Index

Name and Synonym Index

Name and Synonym Index

Name and Synonym Index

Name and Synonym Index

Name and Synonym Index

Name and Synonym Index

Name and Synonym Index

Name and Synonym Index

Name and Synonym Index

Name and Synonym Index

Name and Synonym Index

Name and Synonym Index

9781138717749